The Gordian Knot

The Gordian Knot
Political Gridlock on the Information Highway

W. Russell Neuman
Lee McKnight
Richard Jay Solomon

The MIT Press
Cambridge, Massachusetts
London, England

Second printing, 1998

This book was set in Sabon on the Monotype "Prism Plus" PostScript Imagesetter by Asco Trade Typesetting Ltd., Hong Kong.

Printed and bound in the United States of America.

Library of Congress Cataloging-in-Publication Data

Neuman, W. Russell.
  The Gordian knot : political gridlock on the information highway /
  W. Russell Neuman, Lee McKnight, Richard Jay Solomon.
    p. cm.
  Includes bibliographical references and index.
  ISBN 0-262-14061-6 (alk. paper)
    1. Information technology—Economic aspects—United States.
  2. Information technology—Political aspects—United States.
  3. Internet (Computer network)—Government policy—United States.
  4. Information networks—Government policy—United States.
  I. McKnight, Lee W. II. Solomon, Richard Jay. III. Title.
  HC110.I55N48   1997
  338.4'7004'0973—dc21   97-1580
                                                              CIP

...the enhancement of commerce in communication by wire and radio so as to make available, so far as possible, to all people of the United States, a rapid, efficient, Nation-wide and world-wide wire and radio communication service.

—Communication Act of 1934, U.S. Congress

Contents

Acknowledgments   ix

Introduction   xi

1 Political Gridlock   1

2 The Nature of Networks   45

3 The Network and the State   85

4 Networks and Productivity   113

5 Network Wars: A Pattern Emerges   155

6 Cutting the Knot   247

Notes   265

References   273

Index   307

Acknowledgments

The research for this book was conducted at the MIT Research Program on Communications Policy and was funded by the John and Mary R. Markle Foundation. Further support for related research and analysis was provided by the Advanced Research Projects Agency's Interface Technologies projects at MIT and Tufts University. The MIT Media Laboratory also provided space and support. Significant contributions to earlier versions of the text were provided by Marvin Sirbu, Suzanne Neil, and David Jull.

The inspiration for this book came from our work with the late Ithiel de Sola Pool and his seminal book *Technologies of Freedom*. Helpful comments and criticism were provided by Mitch Kapor, Bill Drake, Linda Garcia, Tony Oettinger, Ben Compaine, Joe Bailey, Sharon Gillett, David Gingold, Loretta Anania, David Tennenhouse, David Clark, David Staelin, William Schreiber, Ellen Staelin, Andy Lippman, Ken Phillips, Tice De Young, Clark Johnson, Jr., Eli Noam, Peter Huber, and Anthony Rutkowski. The editorial assistance of Julia Malik, Gillian Cable-Murphy, and Melita Serdar was invaluable. We also would like to express appreciation to Teresa Ehling, our patient acquisitions editor at The MIT Press, to Jeff Chow, creator of the collage appearing on the cover, and to our colleagues and families for their understanding and support as we struggled to untangle the Gordian Knot—seemingly one strand at a time.

Any errors of fact or interpretation are the authors' sole responsibility. The views expressed here are those of the authors, and should not be taken to represent the views of any of the sponsoring institutions.

Introduction

As legend has it, in ancient Phrygia on the vast plateau between the Black and Mediterranean Seas in what is present-day Turkey, King Gordius tied a large knot of tangled rope. He took special pleasure in challenging visitors to try to untie the Gordian Knot, claiming that he who did would rule all of Asia. Many tried and failed. When Alexander the Great passed through Phrygia, he was of course confronted with the traditional challenge, which he accepted without hesitation. He drew his sword and cut the Gordian Knot with a single dramatic stroke.

In common usage today, the Gordian Knot metaphor is used to identify problems that require a bold stroke and fresh thinking for a successful resolution. Given the frustrating complexities of so many of today's social problems, the metaphor is particularly seductive to the policy analyst and, alas, we have succumbed. It is an unlikely title for our book because, generally speaking, we share a rather strong skepticism of dramatic and simple resolutions proposed for complex political and economic problems. But we have become convinced that in this case the metaphor is apt. We will attempt to persuade our readers that our conviction is well founded and our judgment sound.

The focus of our case study is the entangled policy debate involving swift and unprecedented changes in the economics, control structure, and technology of human communication that we identify as the communications revolution. The historical driver here is technology. We will argue that the generic properties of the evolving communications technologies are well understood. The social, economic, and political ramifications of these changes, however, represent a significant and enduring puzzle. Although far from technological determinists, we believe it is important to acknowledge that the technology of the integrated digital electronic network has inherent properties that distinguish it from the disconnected mix of phonograph records, letters, books, newspapers, and telephone, computer, and broadcasting systems that preceded it. What is subject to our control is how these technologies are structured to meet human ends, and it is in the spirit of that challenge that we proceed.

In selecting a focus for analysis, authors are frequently counseled to pick a historical case of appropriate obscurity rather than a prominent or current event. The inevitable critical reactions of those with vested interests, and the likelihood that new developments will simply overtake the analyst, put the researcher in an awkward position. In ignoring such well-considered advice, we hasten to explain that this book is a bit of an exercise in catharsis.

We characterize ourselves as veterans of the high-definition television (HDTV) wars of the 1980s, a particularly intense national and international competition for technical preeminence in advanced television. Through our association with the Advanced Television Research Program at the Massachusetts Institute of Technology, we conducted a series of studies on audience reactions to new display technologies and a parallel series of economic and policy studies of HDTV. Although from our point of view we were simply conducting the best scientific research possible, we were well aware that we were perceived as being on the "MIT team" and thus supporting MIT's entry into the TV standards sweepstakes. After work we would gather to reflect on the day's events and discuss which company or research laboratory was winning or losing the most recent technical competition. But in these discussions we were repeatedly struck that our fellow researchers seemed oblivious to the terrain of the battlefield on which they were fighting. As social scientists, whose careers would presumably not be judged on the basis of which engineering design won out, we came to define ourselves less as soldiers than as war correspondents and strategists, chroniclers and students of an intriguing and serious techno-economic conflict. Why did so few understand the rules of the game?

We attempt to accomplish three things in this book: explain the current technological debate in terms that a nontechnologist can easily understand; put the technological debate in a broader historical context, focusing on the political economy of technical change; and offer some conclusions on what a policy framework for managing the communications revolution should include, as well as a critique of the recently passed Telecommunications Act of 1996.

This is a book about how technical change is forcing a policy paradigm shift. Since the era of the robber barons, the technical character of telegraphy, telephony, and broadcasting required that each be regulated as a public-service monopoly. The digital revolution (in which HDTV is a relatively minor development) will not require regulated monopoly provisions; rather it will be best served by open, undistorted, healthy economic and technical competition. As we review the literature, we find universal acknowledgment that in a vaguely defined and distant future, competition will be the norm; but until then, there must be new legislation and armies of regulators to manage the transition. This strikes us as a big mistake. The Telecommunications Act of 1996, for example, contains more than 100 pages of detailed legal mandates and prohibitions in an ill-advised attempt to micromanage the transition to deregulation rather than cut the knot. The act calls for more than 90 bureaucratic inquiries, evaluations, and rulings by the Federal Communications Commission as part of its implementation—which, despite the FCC's best intentions, are likely to drag on for years. Some provisions of the act were challenged in court within hours of its being signed into law.

The act will only delay and distort the process unnecessarily. It is best to proceed quickly. It is fruitless to attempt to untie the knot one cord at a time. We term our proposal Open Communications Infrastructure (OCI), and it will be developed in some detail in the pages ahead.

There are two schools of thought in communications regulation, reflecting the polarization of economic theory more broadly defined: one emphasizes the free market, and the other focuses on the need for at least some regulation. Our analysis proposes a third perspective. The conservative free-marketeers emphasize that market competition is the most efficient allocative mechanism. According to this view, the regulators are the bad guys because they afflict the open market with the distortions of political power. More "liberal" policy analysts, however, point out that although markets are indeed generally efficient, there are classes of market failure that require authorities to step in from time to time and substitute their judgments for those of the market gone askew.

The conservatives argue that the regulators, although perhaps well intentioned, are slow to respond to technical change, ill-equipped to assess value and benefit, and inevitably subject to distorting political pressures. We agree. The liberals argue that left to their own devices, the capitalists are likely to be rapacious, monopolistic, and cavalier about the public interest. We agree with this observation as well. As a result, we believe that a new paradigm for regulation is required—one that puts a reformed regulatory structure in the position of ensuring meaningful competition while abandoning its role as arbiter of tariffs and definer of public service and public interest.

The need for a dramatic shift in the policy paradigm is supported by our observation that technical change in this field is proceeding much more rapidly than has been generally acknowledged. Much of what is now considered to be decades away will likely be of economic significance in just a few years. We watched this temporal cognitive distortion again and again in the HDTV wars. Our associate, William Schreiber, obtained funding in the early 1980s from a consortium of American television networks and equipment manufacturers to provide an American alternative to the Japanese model of HiVision. From the beginning, he suggested to his sponsors that advances in digital technology dictated a close look at a digital television system—not just a higher-quality picture. His appeals were repeatedly rebuffed by the industrial engineers as being too academic and likely to delay the development of a practical system; they considered digital television a good idea for new systems a decade or two hence. For eight years industry continued to submit designs for analog HDTV systems until the very last day to file proposals with the Federal Communications Commission. Late that Friday afternoon in June 1991, an unknown new proponent, General Instrument, a cable equipment company working quietly with some of Schreiber's former students on the periphery of the HDTV community, put forward a working prototype of an all-digital system. Why did they propose such a system while the central players did not? Perhaps because they did not know any better—that is, they were not caught up in the prevailing technical ideology that dictated that such a system was too impractical and difficult to be achieved for decades. Within four months, each of the other four consortia abandoned their previous systems and demonstrated all-digital systems of their own. When necessary, in the process of technological development, decades can shrink to months. It is a pattern that is repeating itself frequently in technology development, and if our arguments find some favor, perhaps in policy reform as well.

The burgeoning Internet, the information highway metaphor, and the idea of a National and Global Information Infrastructure are central to our thesis. The metaphor draws attention to the parallels between transportation and communication. Highways and infrastructure are beneficial public entities that facilitate communication and trade. They are tied to national productivity and competitiveness, and although usually built by private contractors, they are generally designed and regulated by public authorities. This public-private tension is critical. In the late stages of the Clinton transition in mid-December 1992, there was a high-profile discussion among business executives and policy advisors in Little Rock, Arkansas. AT&T Chief Executive Officer Robert Allen remarked that the information highway was of great importance to the economic future of the United States, but that it was not a matter for federal involvement: private industry would design and build the highway, and government regulators should simply get out of the way. There was an awkward flash of tension as Vice President-elect Gore jumped up to note that as with the interstate highway system, the private sector might not be able to go it alone; it was an issue simply too important to be left to commercial largesse. President-elect Clinton broke the tension with a joke, and the conference moved on. But the issue to which we shall refer as the Great Debate—that is, the role of the state versus that of private enterprise—remains.

Many adherents of Gore's school of thought draw on federal involvement in the design and funding of the interstate highway system during the 1950s as the historical model for the electronic highway of the next century. For us, however, the most telling historical parallel would be the railroad robber barons of the late nineteenth century. The federal government, through a variety of incentives including generous land grants and subsidies, did its best to encourage private investment in railways in the interest of commercial trade and geographic expansion. Although there is continued debate among economic historians on the importance of rail investment to economic growth, the huge fortunes amassed by Carnegie, Gould, Vanderbilt, Morgan, and Harriman speak eloquently to the question. Once the infrastructure was built and was threatened by devastating competition, the barons colluded to set prices and protect their oligopoly. Farmers and traders who depended on the rail system to transport their produce to market were outraged at the exercise of market power by the barons, and demanded federal intervention. The American route system was largely completed by the 1890s. It was not until 1888 that the Interstate Commerce Commission (ICC), predecessor of the FCC, was established. Furthermore, it was not until the early 1900s that the Congress and the courts finally gave the ICC the legal authority to exercise meaningful oversight. In retrospect, we have at hand a treasure trove of fifty years of economic and legal chaos—flagrant cases of stock manipulation, price rigging, court cases dragging on for decades, even gun battles—as entrepreneurs fought with each other and with the government to win control of huge segments of the system. The result was too dynamic perhaps to be characterized by the currently fashionable term gridlock; perhaps the trench warfare of World War I better captures the spirit of the stalemate of that era.

AT&T's Allen tells Vice President Gore to get out of the way—it sounds right, resonating with politically conservative orthodoxy. But historically that is not the way it plays out. If a regulatory structure can be manipulated for profitable advantage, it will be. Those who benefit and have become most adept at manipulating the regulations wearily assert the need to do away with government regulation altogether, and it is this delicious irony that we propose to sidestep by cutting the knot. Although we should move directly to meaningful competition in the electronic network of networks, it must be with the understanding that, despite their rhetoric, real competition is not what the CEOs have in mind. Thus there is a continuing and important role for government.

The first chapter introduces these central themes and describes the players and stakes. As we write, a carefully negotiated legislative package entitled the Telecommunications Act of 1996 has just been signed into law. Although the legislative rhetoric celebrates competition and deregulation and the act is described by some observers as encouraging open competition among different industry segments, it is in fact a delicately negotiated package of deregulation, partial deregulation, and re-regulation. The key to the act's character is the use of the term "safeguards." Safeguards in the rhetoric of the legislation speaks to rules and procedures to restrict the potential of the vestiges of monopoly abuse until "true competition" comes to pass. Each of the players wants the safeguards to apply to everybody else, and each will work the evolving system with considerable skill to block any action perceived to be contrary to their interests. As usual, the devil is in the details—and there are details aplenty. Some have jokingly termed the new legislation the "Communications Attorneys' and Consultants' Full Employment Act of 1996." In our view such legislation, like the original railroad legislation, will tend to forestall and constrain real competition, perhaps as then, for decades. New legislation or not, our near-term fate is gridlock.

The second chapter develops the argument that digital electronic networks such as the Internet are not like railways and highways, nor are they like their analog electronic forebears in telephony and broadcasting. The generic properties of the new networks render them inherently unfriendly to monopolies, hierarchies, and centralized control. The locus of control migrates from systems operators to users. The movement from single-purpose to multipurpose networks enhances the capacity for competition. The century-old traditions of common carriage and public-trustee regulation become unnecessary burdens, while both regulators and systems operators deny that their expertise in regulatory gamesmanship is no longer relevant.

The third chapter returns to the central question of the appropriate role of the state in building and maintaining public networks. Despite the American predilection for private ownership, the argument that there is scant need for government involvement in the networking business finds little precedent in economic history.

Chapter 4 focuses on the productivity paradox. The information infrastructure, we argue, is the key to international competitiveness in the information age. Networked information systems that work well become the most important potential competitive advantage, although the connection between information technology and productivity, it turns out, is frustratingly difficult to demonstrate.

The fifth chapter analyzes a series of case studies in the recent history of American public communications to illustrate the paradox of the partially regulated and partially competitive markets for communications services. We review the painfully awkward and largely unintended federal trajectory toward deregulation of telecommunications, the parallel battles in radio and television deregulation, the computer network wars, and the continuing battle for control of wireless communications technologies.

In chapter 6 we return to our policy proposal for an Open Communications Infrastructure. We conclude that, based on the economic history and technological and political analysis presented in the preceding chapters, the Gordian Knot must be cut.

1
Political Gridlock

From a media consumer's point of view, everything is in its proper place. To listen to music or the news, there is a radio in the kitchen, beside the bed, and of course in the car. The TV set is in the den, with a smaller one in the bedroom. For a newly released movie, we go to the suburban multiplex. Like clockwork, these movies will appear next in the video store, then on cable, and eventually on broadcast TV. Given the taken-for-granted natural order of the motion picture release cycle, one almost unconsciously calculates whether to go to the theater or wait until a new film is released on video or broadcast on television. Books can be found in the bookstore. Magazines come in the mail, sandwiched between advertising brochures. The newspaper reliably hits the front porch every morning before seven o'clock. The telephones in the kitchen and den stand at the ready. Nobody turns the system off at midnight so that sleep will not be disturbed. These media have become integrated into the rhythm of daily life in the industrialized world. Somehow accepted practices have evolved, dictating when and to whom a phone call is socially acceptable. The newspaper is read over morning coffee, the TV is turned on as soon as the dinner dishes are cleaned up.

The modern newspaper has been with us since the 1830s, the telephone since the 1870s, magazines since the 1890s, radio since the 1920s, TV since the late 1940s. In each case, when a successful new communications technology evolved, it did so by wedging its way into the daily rituals of human life. The popularity of television pushed the radio from the living room to the kitchen and bedroom, but not out of the house. Radio programming evolved from live comedy and drama to music, news, and call-in talk shows. There were only a few economic casualties of this expansion. The advent of broadcasting, for example, put news-reels and newspaper "extras" out of business. But in the end, the norms (and the underlying economics) of public communication have become crystallized and routinized. Each medium plays to its strength. A compact disc is optimized for the reproduction of high-fidelity music. One could have designed the telephone system for delivering broadcast high-fidelity music on demand, but the system instead was optimized for dial-up two-way voice communication.

The digital revolution in communications technology disrupts this relatively stable equilibrium of technology, economics, and human behavior. For the last 150 years, as new communications technologies were invented and introduced to the public, the successful ones were those that elbowed their way to a place at the table. To protect themselves, they became differentiated. For the most part, they created de facto monopolies and reaped spectacular profits.

The communications revolution, however, has set off a technical implosion. Through digital processing, a single medium can offer all the services once provided by a range of media. Packaged and print media can move to electronic delivery; the telephone company can deliver multichannel television; the cable company can provide telephone service; and each of these formerly distinct services (along with other competitors) can provide electronic home shopping, electronic encyclopedias, magazines and newspapers—all delivered to high-speed home printers. What were once noncompeting, parallel, and highly profitable industrial sectors of the economy have all been thrown into the same electronic marketplace. The cash flow of this new industry will probably significantly exceed the combined totals of their individual predecessors. New forms of transaction and subscription services on top of traditional information and entertainment media will fuel this expansion.

Some observers predict that there will be a political and economic battle of unprecedented intensity among the giants of the global media, information technology, and telecommunications industries. International megacorporations like AT&T, IBM, Time-Warner, Disney, and News Corp are the hard-nosed, deep-pocketed survivors of previous battles in which they came to dominate one or more separate media markets. The high-power politics of jockeying for strategic position has already started in earnest. Publicly, media executives are quick to express enthusiasm for the new electronic superhighway, but behind closed doors they acknowledge that the media revolution is a potential economic disaster they would have done well to avoid. Consumers meanwhile are a bit puzzled. They squint at the information highway, which to them looks suspiciously littered with hyped-up technobabble. Many suspect that emerging technologies will make media costs go up rather than down, widening the gap between the information rich and the information poor.

Most consumers lose interest quickly when asked whether their television program came over analog or digital lines. Government officials are concerned, however, because they know that decisions about technical architecture translate into economic and political winners and losers. They anticipate the upcoming battle royal, but their role is unclear. Cautious politicians seek cover from the pressures of powerful forces, while a few brave ones pursue leadership roles to hammer out compromise. On the other hand, academics who study this field are ecstatic, because the communications turf war promises to offer one of the most interesting case studies of high-powered political economic conflict ever— that is, the opportunity of an academic lifetime.

Our intent in this chapter is to introduce the key issues and themes surrounding the transformation of the political and economic life of the nation arising from the diffusion and growth of digital computing and communications systems. Some readers may be disappointed that we largely ignore the fiery debates and ferocious turf wars that result as the communications titans struggle to develop future markets while defending existing ones. Instead we present a long-term view of the evolution of information and communications technologies and services over the centuries, and draw from this picure what we hope is a simple conclusion: the time has come for an Open Communications Infrastructure. As we envision it, OCI promises to eliminate the political gridlock affecting information and communications technologies, and—if we may break our promise not to indulge in the political rhetoric of the moment—it will speed us along on the information highway toward the information-intensive society of the twenty-first century.

First we will introduce the players vying for control of the information and communications systems and the playing field of economic conflict where gridlock is one of the most important weapons. We then address the question of global competitiveness, which turns out to be rather difficult to assess in real time as we hurtle through cyberspace and as economic activity becomes increasingly difficult to measure using traditional tools. We take up the issue of whether the United States is facing continued stagnation in its standard of living on an information highway paved by low-wage workers, or conversely, whether the nation has dynamically restructured industry and raised productivity to enable future economic growth. These issues will be addressed in later chapters. First, the clash of the titans—now playing in the halls of Congress, in the courts, in the marketplace, and in the media (if not at the local cinema)—will be reviewed.

The Clash of the Titans

Historical Precedent in Economic Conflict

Economic historians may counsel less scholarly enthusiasm and more hard-edged realism regarding the process of economic change: Large-scale economic turf wars, they like to point out, are in more than ample supply. In U.S. history the festering tensions in the industrial North and the agricultural South culminated in the Civil War, which some claim was as much about agricultural economic independence as it was about slavery (Walton 1994). Following the war, the westward expansion was marked by a battle between two agricultural sectors, cattle ranchers and farmers (Degler 1967). In the 1890s, the battle lines shifted to monetary policy. The conflict between forces in favor of maintaining a gold standard and those advocating the free coinage of silver was settled by the election of 1896 (Friedman 1963). The heat generated by this dispute, however, far exceeded the ultimate economic importance of its resolution in favor of a gold standard.

The nineteenth century witnessed technological change in the nature of networks, with the transition from roads and canals to railroads. After the turn of the century, technological change continued, from railroads to motorized trucking and ultimately air transport. In communications there was the introduction of the telegraph in 1844, and fifty years later the transition from telegraph to telephone. These developments brought into being the era of the robber barons. The railroads were the first large-scale exemplar of corporate capitalism. The centralized economic power of these new corporate entities was unprecedented, and the capacity of the government to respond to abuses and exercise legislative oversight predictably lagged for several decades. Significant federal legislation affecting the nature and future structure of U.S. transportation and communications networks took a long time to emerge. This legislation included the Interstate Commerce Act of 1887 and the Sherman Anti-Trust Act of 1890, which attempted to define the problem and establish the jurisdiction of the federal government in regulating such infrastructure (among other things), but which mandated little power of enforcement. The Mann-Elkins Act of 1910 and Clayton Act of 1914 after a century of laissez-faire policies finally put some teeth into federal enforcement. Realpolitik legislation during the 1920s countenanced AT&T's de facto hegemony for the next sixty years, but left sufficient grounds for continual antitrust challenges, culminating in the 1982 breakup.

In the 1990s, we are experiencing technical change in communications and transportation at a pace similar to that of the middle-to late nineteenth century. The wild era of post-Civil War political economy rewarded aggressive and quick-witted economic entrepreneurship. This predictably led to abuses during the period of economic restructuring. Federal authorities only belatedly recognized the dramatic character of change, and responded still more slowly to the resulting political and economic distortions. We expect that drama to repeat itself into the twenty-first century: same play, new characters, modern dress.

In the following chapters we develop this argument and propose a historically self-conscious approach for appropriate, balanced, and realistic public policy in support of an Open Communications Infrastructure. But first, the players and the play should be formally introduced.

The Players

The would-be robber barons of the 1990s have considerable financial resources with which to expand and protect their empires. Table 1.1 roughly outlines the industrial sectors as they stand now, on the eve of convergence. The core industries represent about $400 billion a year in

Table 1.1
Converging industrial sectors of the $400 billion information economy: Revenues preconvergence (1993 figures in $ billions)

Industrial sector		
$134	Telecommunications	
	Local	$83
	Long distance	51
124	Publishing	
	Newspapers	45
	Direct mail	27
	Magazines	22
	Books	18
	Information services	12
63	Broadcasting	
	Television	27
	Cable	25
	Radio	10
	Satellite	1
53	Computers and data networks	
	Hardware	28
	Software	20
	Data networks	5
20	Consumer electronics	
5	Theatrical motion pictures	

Source: Weller and Hingorani 1994.

gross revenues in the United States alone. As two-way electronics allows penetration into the neighboring turf of retail and information services, the estimate is expected to approach $1 trillion a year—roughly 20 percent of the U.S. gross domestic product (GDP). The other 80 percent of the economy is directly affected by the technical and market changes in information and communications systems, as we see for example when businesses turn to information technology in their effort to engineer new efficiencies into their manufacturing cycles. It is not difficult to see why corporate boards, CEOs, and strategic planners are attracted to the idea of having a significant portion of this enormous pie become their proprietary electronic real estate.

The Playing Field

There is a significant difference between the era of Gould, Morgan, and Carnegie and the present day. For most of the 1800s there was no federal regulatory tradition. Now there is not only a tradition, but a large number of turf-conscious bureaucracies and a comprehensive ideology of government oversight covering even the most far-flung communications industries. It may prove easier to invent new public-policy instruments than to reinvent or redirect existing ones.

In the current policy debate one frequently hears a plea for a level playing field. All we want, telephone, cable, and publishing executives are wont to claim, is to tip the playing field back to level, to correct the imbalance that currently favors the other guy. There are several problems with such a metaphor. First, there are dozens of industrial sectors involved in this political struggle. It is hard to imagine a field tipped in twelve directions at once. Second, the direction of the tip depends on where you stand. As with the greener grass on the other side of the fence, each player eyes competitors' turf with special anxiety. It is not just self-serving cynicism; the combatants come to view these inequities with both considerable emotion and sincerity.

Perhaps a better metaphor is the Ptolemaic model of heavenly motion. As our ability to measure the movement of the sun, the planets, and the stars improved, medieval astronomers stuck to their geocentric model, but they invented little subtheories to explain the anomalies. They observed brief periods of retrograde motion and posited that planets moved not in a given orbit but in a tiny suborbit around an invisible object in the original orbit. After a while there were orbits around orbits around orbits—an unbelievably unwieldy and unlikely scenario. An astute astronomer might have sensed that the time was right for a paradigm shift.

Ultimately the heliocentric model of planetary motion resolved this problem and reintroduced a model of striking and refreshing scientific simplicity. We will make such an argument with regard to the current state of telecommunications regulation and the need for creating an Open Communications Infrastructure as the new paradigm.

For our current purposes, however, we want only to highlight the Ptolemaic character of the current system. After a century of laws written

(a)

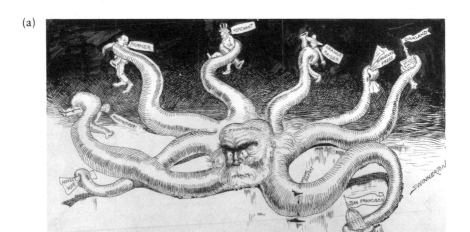

(b)

to regulate grain elevators and railroads, which were edited to incorporate telegraphy, telephone, radio, television, satellites, and computers, then further embellished by a litigious century of accumulating case law, we have an awkwardly uneven legal playing field of Ptolemaic complexity—a playing field only a lawyer could love.

The accumulated inequities of the playing field are easy to demonstrate. Table 1.2 illustrates several examples of the current array of regulatory distinctions, ranging from subsidies to a variety of circumscriptions. These regulatory distinctions for the most part reflect the logic of the moment when these industries first emerged as identifiable business sectors. The earliest example was the post office. It made sense for the government to both provide service and encourage the free flow of information and a national sense of community, in spite of physical isolation across a sparsely populated continent, by subsidizing the conveyance of newspapers, books, and magazines. The early days of newspapers were marked by robust competition, with over a dozen dailies available in many larger cities. There was no need for regulation in such a market; indeed, the political tradition of the independence of the fourth estate and the First Amendment itself would seem to preclude it.

Broadcasting is rate regulated, with provisions that guarantee the lowest possible advertising rates for political ads; and content regulated through prohibition of "offensive" content and requirements for children's programming. The regulation of radio and television broadcasting content was based on the "public trustee" principle. The rationale was that since the public airwaves were provided to competing commercial broadcasters without charge, there was nevertheless a public-trustee obligation to abide by publicly determined content guidelines.

Figure 1.1
The demonization of corporate power. Whether (a) Collis P. Huntington's Southern Pacific Railroad, or (b) Bill Gates's Microsoft Corporation, neither the press nor the public tend to express warm feelings about large corporations. The octopus theme persists. One might predict an over time correlation between octopus cartoons and an energized antitrust division within the Department of Justice. (a) Courtesy of the California History Room, California State Library, Sacramento, California. (b) Copyright © 1995 by the New York Times Company. Reprinted by permission.

Table 1.2
Regulatory models of information industries: Government provision, subsidy, rate, and content regulation

	Provision	Subsidy	Rate reg.	Content reg.
Publishing				
Newspapers		X		
Magazines		X		
Books		X		
Recorded music		X		
Motion pictures				
Broadcasting				
Radio		X		X
Television		X		X
Cable			X	
Satellite				
Advertising				
Media				
Promotion				
Direct mail		X		
Information services				
Consumer electronics				
Postal services	X			
Telecommunication			X	
Computation				
Hardware				
Software				

Broadcasters who violate the content rules can be reprimanded, fined, and ultimately have their broadcast license revoked.

Cable television systems were originally not regulated at all at the federal level because they do not use the radio spectrum. As their market penetration and economic importance rose in the 1960s, the Federal Communications Commission, at the behest of the broadcasting industry, simply declared that cable was "ancillary" to broadcasting and thus subject to FCC regulation—a declaration of authority ultimately sustained by the courts (Kellogg, Thorne, and Huber 1992). Given that cable service, unlike broadcasting, requires access to utility poles, cable companies are subject to contractual and regulatory constraints from local and state authorities. This local authority, however, was severely restricted by the Cable Act of 1984. Through the act the increasingly wealthy cable industry gained so much power and independence from competition and rate regulation that the resultant rise in cable rates generated one of the most curious legislative turnabouts—the Cable Act of 1992, which attempted to roll back and reregulate cable rates. In this curious political history nobody got around to regulating cable's potentially offensive content. The logic was that broadcasting through the airwaves is somehow "invasive" and should be available to all without charge. Cablecasting requires the customer to proactively order and pay for the service, and thus one can simply disconnect it in the event of perceived offense.

Because telephone service was judged to be a natural monopoly, the government agreed in 1913 to forbid new competition against AT&T. In return AT&T accepted federal rate regulation as a guard against the prospect of monopolistic tariffs. With the tradition of common carriage, however, telephone carriers, unlike broadcasters, were not responsible for the content of what was carried over the system. This precludes the use of content-oriented regulation. The need for telephonic content regulation started to gain supporters in the 1980s as commercial adult-oriented services sprang up, and both federal and local regulators decided to promulgate rules for restricted access to these kinds of services.

The absurdity of the current system can be illustrated by examining the possible regulatory constraints on a single message as it might pass through different media of mass communication. For example, established copyright laws work well with the established print model, but they break down with electronic publishing. Computers can alter text in an electronic form through a simple computer program, but it is unclear if a computer is guilty of copyright infringement. Can a computer author copyrighted material (Pool 1983)? It turns out that whether the message is transmitted electronically or in print matters a great deal to the courts and regulators. It may be of little consequence, however, to the sender or receiver, except for the regulatory impediments. Eliminating this uneven playing field, which distorts and inhibits communication, is a principal reason for adopting the Open Communications Infrastructure regulatory model.

The Rhetoric of Gridlock

We have witnessed an interesting transition in the rhetoric of competition among the communications industries over the last several decades. We identify three stages.

The first stage, spanning much of the first half of the twentieth century, might be characterized as one of sectoral independence. Each communications industry had its own turf, and the boundaries were clear-cut. There was some competition across sectors—for example among publishers, broadcasters, and telephone yellow pages for advertising revenues—but the focus of competition was within each sector (magazines versus magazines, one radio station versus another). The measures of industrial success and failure were industry specific—Nielsen ratings, circulation figures, telephone penetration. The rhetoric was low-key, even polite.

The second stage emerged in the 1980s, reflecting a growing recognition of the prospects of meaningful cross-sectoral conflict and competition. It was also characterized by a rush to the barricades. The impulse was to protect sectoral boundary lines. One of the most intriguing incidents was the standoff between the newspaper and the telephone industries. Newspapers derive 30 percent of their income from the sale of classified advertising, and they recognized that dial-up information services and computer bulletin boards threatened to make significant inroads into this income stream. The American Newspaper Publishers Association (now renamed the Newspaper Association of America) set up high-level committees to plot strategy and lobby Congress for protection. Basically, they argued that permitting telephone companies to enter the information service business would put newspapers out of business. Similarly, the Regional Bell Operating Companies (RBOCs) were to be kept out of long-distance and manufacturing businesses for fear of the negative consequences they would produce for their competitors. The political maneuvering was intense and the rhetoric zero-sum, but both remained largely out of public view.

The third stage, the battle of survival, follows a change in strategy in which the sectoral players acknowledged that there was going to be a technological shift and that an advanced electronic network would be built. But now each sector claims that it alone is qualified to build it. Each claims that there is no need for the competitor's network. Each claims to have a better technological platform to get the job done, and each also claims that the other will function as a monopolist. The telephone company, the cable company, and the computer-network company each imagine a single high-capacity network in the future, connecting every firm and every household. It would be wasteful, they argue, to build more than one network. It is true, of course, that advanced technology permits virtually all of the services under discussion to be provided by a single network. The battle for survival is over who will ultimately own this proposed single, universal, high-capacity network, forcing all the others to do business on its turf. Because each of the major players dreams of owning this great engine of economic power, they are all inclined to argue for the "special efficiencies" of the single-network model, while at the same time warning of the evil intentions of the others.

Examples abound. Robert Kavner, then executive vice president of AT&T, gave a keynote address at the 1994 consumer electronics industry trade show. Stay away from the cable industry for the provision of interactive services, he warned. "They will act as gatekeepers, restricting what goes into homes over the emerging information highway. Working with the cable industry is like picnicking with a tiger; you might enjoy the meal, but the tiger always eats last. The cable industry threatens the very existence of the consumer electronics industry because of cable's stranglehold control on the set top box" (Carlson 1994). A month later the cable industry filed an FCC petition asking that Pacific Telesis (PacTel) be forbidden from offering advanced services on its proposed $16 billion high-speed video and data superhighways. Why? Because the construction would be unfairly subsidized by telephone customers paying higher phone bills. PacTel responded, "They are trying to use the regulatory process to impede competition in their marketplace, at the same time they are trying to get into our marketplace." "This is about survival," they added (Adelson 1994).

The survival of many firms in a large number of information industries is now threatened as market share, technology, service design, and network architecture shift seemingly minute to minute. New cross-sectoral research, standards, production, distribution, and marketing alliances are announced daily, while others are renounced quietly. And the battle does not stop at the water's edge; rather, it is fueled by global communication links.

The War's Second Front

Thus far we have characterized the clash of the titans primarily as a battle among domestic industrial groups that seek economic dominance. Still, we live in an age of increasingly interpenetrated national economies and multinational corporations. Goods, services, and people cross international boundaries with increasing ease, speed, and frequency. Economic integration in Europe, Asia, and the Americas takes on new significance. The communications revolution is both a central cause of globalized economics and the central playing field on which economic competition among nations takes place. In the information age, if one nation or region manages to attain significant control of the electronic playing field, it will have achieved a remarkable competitive advantage. Such questions attract considerable attention in Europe, Asia, and the Americas. Analysts in each region alternatively fantasize about the prospect of their own success in seizing dominance, and fret over the dire possibility that others will.

In 1988, Richard Munro, the chief executive officer of Time Incorporated, pondered the future as the unprecedented Time-Warner merger was being finalized. In one public statement he mused, "By the end of the decade there will probably be only four or five dominant media conglomerates" (Bagdikian 1989). This is not the sort of thing one wants to draw to the attention of the Justice Department, but his candor is refreshing.

American Chaos versus Coordinated Efforts in Japan and Europe

Concerns about the American communications infrastructure are deepened by the realization that extraordinarily well-financed, coordinated plans for system enhancement are being undertaken in Europe and Asia. As early as 1972, the Japanese government circulated its report, *The Information Society: A Year 2000 Japanese National Goal.* It outlined a stepwise plan of advanced techniques for manufacturing semiconductors and building national data networks and database technologies. In 1981, Nippon Telegraph and Telephone (NTT) put forward a plan for an information network system (INS). In 1984, the Ministry of International Trade and Industry (MITI) announced a complementary plan for advanced communications systems in model cities. In 1985, the New Telecommunications Law established a dramatic new process for deregulatory reform and system development. In 1987, the Private Sector Vitality Act provided additional tax preferences and interest-free loans for telecommunications infrastructure development and research projects. These coordinated policies and their recent revisions are expected to result in a $250 billion investment in Japanese infrastructural development. Many other projects have sought to improve Japanese competitiveness in related electronics industries, such as semiconductors, and in some cases they have succeeded

The European situation is similar. In this case the seminal document is the French government's *Nora-Minc Report* of 1980, which called for massive enhancement of the French network by expanding telephone penetration, the highest percentage of digital switching, and the largest packet-switched network in the world. There were parallel projects in England, which included the establishment of a new independent government agency, OFTEL, whose mission is to promote advanced telecommunications development. In Germany, the Witte Report, *Restructuring of the Telecommunications System,* was released in 1987; it was followed by the restructuring of the Bundespost and the opening of a number of markets to competition. Plans for privatization of Deutsche Telekom are progressing. By 1997, the most modern network in the world will be installed throughout the former East Germany, with the hope that this advanced information infrastructure will rapidly raise the living standards and the quality of life of people in the region.

These national initiatives are coordinated with European Union (EU) efforts including: the European Strategic Programme for Research in Information Technology (ESPRIT); Advanced Communications Technologies and Services (ACTS, formerly known as RACE); and the industry-initiated EUREKA program for advanced technologies. These projects have yet to demonstrate their effectiveness in bringing new products to market through the harmonization of European standards and cross-border research and development collaboration.

Global Competition and American Decline?

The concern over the state of the United States' physical and electronic infrastructure has been drawn into a broader debate between two schools of academic internationalists—the "declinists" and the "traditionalists." Books such as: *The Reckoning* (Halberstam 1987), *Trading Places: How We Allowed Japan to Take the Lead* (Prestowitz 1988), and *The End of the American Century* (Schlosstein 1990) have been written by the declinists. The traditionalists have responded with *The Myth of America's Decline* (Nau 1990) and *Born to Lead* (Nye 1990). There is agreement that the U.S. position in technology, trade, and economic productivity declined relative to that of Europe and Japan; but the real debate is whether the problem justifies major changes in the way the United States runs its economy and polity, or is a result of a U.S. industrial restructuring leading to resurgent competitiveness.

Let us take a look at the numbers. The U.S. trade deficit has indeed grown, and the United States went from being the largest creditor to the largest debtor nation in the world, reversing sixty years of investment in only four years. Despite attempts from the administration and Congress to respond to the challenge, the trade deficit continues to accumulate at a rate of about $100 billion per year (Krugman 1990).

The U.S. share of world manufacturing fell from 33 percent in 1950 to about 23 percent in the 1980s and 1990s. Productivity growth, which held strong from 1900 to 1970 at an average of 2.3 percent (2.8 percent in the 1950s and 1960s) fell precipitously to 1.2 percent in the 1970s and 1980s, while preliminary data for the 1990s looks more encouraging. If the United States had been able to maintain the original rate of productivity growth through the decades, living standards for American workers would be fully 25 percent higher than the levels that have actually been achieved. In fact, measured in real dollars, the wages of American workers have fallen since the 1970s, putting economist Paul Krugman in a position to note with some irony that "the last real increase in wages came at about the time of Nixon's first inaugural" (Krugman 1990). Workers in Japan and Europe during the same period experienced demonstrable increases in their standard of living. Such patterns are no mystery. The fundamental economics tell the story. U.S. manufacturing productivity growth from 1960 to 1986 was only about half that of Japan, Germany, France, and Italy, and below that of England and Canada (Krugman 1990; Nye 1990; Office of Technology Assessment 1988). In the 1990s, early indications of renewed U.S. productivity growth appeared, but it is too soon to say whether a reversal in national productivity growth trends will be sustained.

The MIT Commission on Industrial Productivity (Dertouzos 1988) studied manufacturing and trade from 1971 to 1987. The commission documented the decline and noticed that by the end of this period, a significant negative balance of trade existed in all but two. Even though the United States has continued to perform well in software and services, many consider the semiconductor to be the engine of the information age. The decline of American participation in memory chips is among the most dramatic. Prestowitz (1988), for example, reports that in only seven years (from 1980 to 1986) the United States and Japan literally traded places. In 1980, the United States had 75 percent of the market, whereas Japan had 25 percent. By 1986, Japan had 65 percent, the United States had 28 percent, with Europe and the Pacific rim manufacturers holding the balance. For memory chips, Japan dominated 92 percent of the market (Ferguson 1988) although Intel and other U.S. producers dominate high value-added markets for central processors and other chips. Perhaps the most significant fact of all is that national per capita income in the United States has been stagnant since 1973. Although we do not want to pursue this further here, it is worth keeping in mind as we consider how to move the nation forward with new rules of the game for the information economy.

The traditionalists, for the most part, do not deny the numbers; for them the difference is in the interpretation. They are concerned that an overreaction to economic cycles and the emotionally charged term of "decline" itself will lead to the premature abandonment of sound policy principles of free trade and private-sector-dominated economic management. We agree that if the perception of decline leads to narrow protectionism, abandonment of international responsibilities, and a reflexive propping up of manufacturers in economic difficulty, such ill-considered responses would indeed hasten further decline. We are equally persuaded, however, that to coast along confidently without carefully examining strategic options in light of developments would be equally ill-advised.

One of the central themes of much of this literature is that the communications, computation, and information industries will be at the forefront of the battle for international economic predominance. It turns out that the relationship of information technology to industrial productivity is highly complex—and one we will address in a later chapter. Our present concern is the evolution of the U.S. policy position vis-à-vis the relationship of information technology and economic health. It turns out that two dramatically different positions have been taken—the laissez-faire approach of the Reagan-Bush era and the National Information Infrastructure initiative introduced by Clinton and Gore.

The Evolution of U.S. Communications Policy

Our particular interest is on the role of the information infrastructure in the communications age. The position of this infrastructure is central to the nation's development. The analysis here points out the historical parallels between the expansion of the canals, railroads, and highway systems in the industrial revolution in the United States, and the need for a comparable public network of electronics and optics today. The National Telecommunications and Information Administration (NTIA), for example, in introducing its comprehensive overview of this issue, notes that:

The U.S. economy is becoming increasingly dependent on the provision of services that require efficient distribution and dissemination of information. Over 50 percent of all U.S. workers are currently employed in information-intensive service industries that are heavily reliant on telecommunications. Even traditional manufacturing firms increasingly depend on the swift movement of information from headquarters to factories to distribution points to customers in order to remain competitive with their domestic and foreign rivals. This transformation has heightened the importance of telecommunications to the nation's economic and social welfare. Many now believe that the telecommunications infrastructure will be as important in the future as the transportation infrastructure has been to the industrial economy. (National Telecommunications and Information Administration 1990, p. 5)

There is a concern, however, that our telecommunications infrastructure is less than it should be.

The Davidson-Selwyn Debates

William Davidson of the University of Southern California filed reports with the FCC and NTIA documenting that "while the United States has been a leader in many aspects of public telecommunications, its lead has narrowed or disappeared altogether in a number of critical areas." He noted that the United States trails France, Canada, the United Kingdom, and even Hong Kong and Singapore in digital switching of local-exchange service. He found that total U.S. investment in public infrastructure actually declined in the late 1980s, and that levels of sustained infrastructural investment in the United Kingdom, Japan, Switzerland, and Germany exceed U.S. levels by 50 percent to 150 percent (1990a). His work is supported by the Regional Bell Operating Companies and is frequently cited as evidence of the need for regulatory relief for the public-switched network.

Veteran analyst Lee Selwyn and his colleagues at Economics and Technology, Incorporated, writing for the International Communications Association and representing the telecommunications user community, are skeptical. They utilize different measures of system performance and different exchange rates for the calculation of relative infrastructure development and conclude that the United States remains far ahead of the pack. Davidson responds that their numbers are out of date, contending that they ignore the downward trend and that Selwyn's exchange rate calculations distort the national comparisons (Montgomery, Selwyn, and Keller 1990).

We cite this controversy not to debate the technical merits of alternative measures of system performance, but merely to note that the issue of the relative strength of the U.S. telecommunications infrastructure is debatable and not a foregone conclusion.

American Policy Paralysis

How can U.S. industry and the nation's citizens respond to these challenges? Should a partially revised law dating from the 1930s continue as the guiding policy statement for communications into the 1990s, given the policy paralysis we currently face? Is the Telecommunications Act of 1996 a sufficiently radical reform to move the United States beyond policy gridlock?

Stalemate at a New Level

As a worst-case scenario, we may find that the telcos, broadcasters, cable industry, private radio and consumer electronics manufacturers, computer companies, and publishers, although they all have their eyes on their neighbors' markets, opt to stick with what they know best and protect their cash flow, sunk costs, and existing plant and choose to challenge and undercut the deal of 1996. In the wake of the Telecommunications Act of 1996, litigiousness is rampant, and ingenious new forms of legal foot-dragging are invented. Meanwhile, the administration and Capitol Hill continue to be paralyzed by intense cross-pressures.

Under this scenario, new technical standards committees proliferate as their function shifts from setting standards to blocking them. Each vendor continues to believe that it can outmaneuver all the others into adopting its own proprietary technology as the accepted technical standard for interconnection, and to that end resists cooperative efforts. The FCC bows to industry pressure and attempts to conduct technical tests to pick the best technology based on "objective technical criteria." The testing process is conducted and controlled by industry consortia because the FCC lacks funding. The tests take forever, and the test results are ambiguous. Half of the vendors take the FCC to court to demonstrate that the tests are unfair. None of the tests has any relevance anyway, because by the time the testing is completed, a new generation of technology has been developed.

In a less stark scenario, we may find that the 1996 act provides for a modest step forward and, despite the efforts of some industry stalwarts, the technology prevails. The local-exchange companies lose their battle against bypass. Local television and radio broadcasters lose their local monopolies as satellite and terrestrial spectrum, coaxial, twisted pair, and fiber wireline each come to offer equivalent services.

There are three problems with this version of the future. First, it significantly delays the development of an advanced, programmable network, with the attendant disadvantages to economic efficiency and productivity. Second, it builds an interoperable network out of pieces explicitly designed not to work that way, leading to inefficiencies and operational awkwardness. Third, and perhaps most important, policy gridlock threatens to hold U.S. companies back from the cutting edge, and effectively abandons the field of integrated communications to Europe and Japan.

The Japanese, on the other hand, have a well-integrated game plan. They have harnessed the Japanese *zaibatsu* tradition, involving a complex pattern of intercompany cooperation and competition, with dramatic success. In Japanese culture, it is a strongly held belief that because the nation is small and relatively poor in natural resources, it must have a highly coordinated and aggressive strategy for economic survival in an increasingly competitive global economy.

The Europeans also have a game plan. It is even more complex than Japan's because it involves the painful economic reorganization of a weak, fragmented array of countries into the largest, most powerful unified economic and political entity in world history. It had a rocky start and has taken decades to develop, but the direction of European integration is no longer questioned. However different the Japanese and European approaches are, one thing is clear: Both heavily emphasize the strategic importance of communications infrastructure to economic productivity.

Does the United States have a game plan? Throughout the Reagan and Bush administrations, the traditional U.S. aversion to industrial policy led to a pronounced and public disavowal of any such thinking (Seib 1990). Behind the scenes, however, there was active discussion between the public and private sectors as white papers were drafted and conferences organized. The format reflected the traditionally stealthy U.S.

industrial policy (McKnight and Neuman 1995), which positions issues of industrial organization and capacity under the rubric of defense R&D. During this period, the National Research and Education Network (NREN) initiative and the High Performance Computing and Communications (HPCC) program, focusing on defense-oriented computer communication, laid the groundwork for a full-scale game plan that was to evolve under Vice President Gore's leadership in the Clinton-Gore administration.

The kickoff for the new initiative took place on 15 September 1993 with a presidential executive order establishing the United States Advisory Council on the National Information Infrastructure and the publication of *The National Information Infrastructure: Agenda for Action* by an existing administration task force. The agenda focused on nine goals: (1) promote private sector investment via tax and regulatory policies, (2) extend the "universal service" concept to ensure that information resources are available to all at affordable prices, (3) act as a catalyst to promote technological innovation and new applications, (4) promote seamless and interactive user-driven operation of the NII, (5) ensure information security and network reliability, (6) improve management of the radio frequency spectrum, (7) protect intellectual property rights, (8) coordinate NII operations with other levels of government and with other nations, and (9) provide access to government information and improve government procurement policies for telecommunications and information services and equipment in order to promote important technical developments for the NII and provide attractive incentives for the private sector to contribute to NII development.

The atmospherics are right. The emphasis on health and educational applications reflect traditional government concerns. The emphasis on promoting commerce and job creation build on the success of the Interstate Highway Act and the Information Superhighway metaphor. Tied in with the rhetoric and publicly announced intent of the Telecommunications Act of 1996, the game plan should be complete and effectively positioned. But we remain concerned. The NII initiative is retreating to the background as the presidential campaign of 1996 takes center stage and the Telecommunications Act threatens to result in more regulation rather than less.

This is not to minimize the importance of the Clinton administration's leadership in initiating public discussion (and cross-agency dialogue) on information infrastructure issues. The administration's initiative energized industry and motivated governments worldwide, as well as at the state and local level, to consider how new networking technologies and services might affect them and their constituencies. Combined with the federal re-engineering initiative, the National Performance Review (also chaired by Al Gore) may lead to a more responsive, efficient, and connected federal government, which in turn may be a legacy of the Clinton administration's National Information Infrastructure initiative.

A Case Study in Political Economics

Political economy is the interdisciplinary study of the interaction of political and economic systems. Much of the work in this field focuses on how market failures of various sorts may require explicit intervention of the polity to protect the environment, provide a social-welfare safety net, constrain monopolistic practices, and so on. From another direction, scholars may study how economic incentives and influences distort an otherwise open process of political decision-making.

This book is a study of a historical transition currently in process as an array of communications industries, once largely regulated and monopolistic in structure, move into an era of deregulation and competition. It represents rich soil for the political economist. One can draw on an extensive theoretical base and a diverse collection of historical parallels and contrasts to better understand the current policy debate. One can also contribute to the developing theory by means of a close examination of a particularly dramatic case. We will try to do a little of each.

To put our position in perspective, we draw on the vocabulary of international relations, which organizes much of its scholarship over the last century around the polarities of idealism and realism. Briefly summarized, the idealists emphasize the potential of extranational institutions and international law to promote international cooperation and reduce the likelihood of resorting to military means for resolving differences among nations. In contrast, the realists—a somewhat more cynical lot—argue that international conflict is fundamentally inevitable, and that a naïve faith in well-intentioned diplomacy is likely to be counter-productive and may lead to misperceptions about the motives and intentions of likely adversaries.

The industrial policy debate can best be understood through the lens of these polarities. Both the free-marketeers and the industrial policymakers have mirror-image realist and idealist world views. Free-marketeers have an unshakable faith in the capacity of marketplace mechanisms to allocate resources efficiently, and they have an equally unshakable cynicism about the ability of government experts to pick winners and allocate resources. Industrial policymakers, on the other hand, see themselves as realists in their view of the marketplace. They consider industrialists astute profiteers who would distort the marketplace to maximize profit and restrict competition, were they permitted to do so. The job of government is to hold these players in check, according to this view, and industrial policy advocates exhibit faith and enthusiasm about the ability of government agents to effectively perform that role.

We characterize ourselves as being similar to the neorealists in outlook: we are deeply cynical students as a result of our study of this dramatic transformation of the communications industry and its roots in the last century and a half of government-industry interactions. We have little faith, however, in either would-be regulators or would-be industrialists to lead us to the information highway promised land. Put bluntly, given the opportunity, both would screw it up.

Regulatory systems work best when the industrial and technological structure is stable, the inputs and outputs easily measured, and the public welfare goals clearly identified. None of these properties is evident in this case. For their part, markets work best when competitive entry and exit is easy, competitive services are straightforwardly comparable, costs and prices are evident and comparable, scale and scope economies are limited, and buyer and seller have equivalent expertise. None of these properties is evident here.

Furthermore, the likely compromise of a half-regulated, half-competitive system of communications networks may actually be worse than either by itself. In our view, many proposals for reform threaten to absorb the worst of both worlds as the economic distortions of the regulatory process and the regulatory distortions of the competitive process diminish the effectiveness of both, while dramatically slowing the pace and reducing the incentives for technical progress. We use the word "halfway" to characterize the mixed nature of these proposals; most of the proposals and analyses we have seen thus far appear to fall into the halfway trap. Halfway proposals by design reach but halfway to their conflicting goals. They are trying mightily to untie the Gordian Knot one strand at a time, and it will never work.

The policy gridlock must be addressed dramatically by means of a clear and distinct policy initiative. Such an initiative posits that regulators are ill-equipped to micromanage such fast-moving technical developments. It also posits that industrialists, who for the most part have enjoyed a century and a half of near-monopolistic profitability, will do everything they can to recapture and monopolize the new network structure. Some media executives seem to perceive it as something akin to a natural right, others perhaps simply as sound business practice. In any case, the laissez-faire approach of leaving the transition to market forces alone is a recipe for disaster.

There is a role for proactive government involvement in managing the transition from a regulated system to one that is truly competitive. But policymakers must abandon the regulations on entry and exit, on prices, and on mandated services at the outset, and focus on nurturing and protecting meaningful competition.

Open Communications Infrastructure

Because this study draws together historical and economic analysis and presents explicit advocacy of a particular position in the communications policy domain, it seems appropriate to signal where we are headed at the outset. In the spirit of the communications field, where power is measured by the obscurity of one's acronyms, we feel compelled to develop an acronym that captures the character of our argument. In public presentations over the last several years we have come to rely on the phrase "Open Communications Infrastructure." "Open" captures the spirit of our call to move quickly away from a regulated common-carriage system to a competitive, privatized, interconnected system of systems. Our proposal would reflect a traditional and conservative slant if we went no further. But "infrastructure" emphasizes the public character of this evolving network and the need for close public scrutiny, proactive government involvement, and safeguards for protection against market failure and the underallocation of resources for public goods. The term "communications" in the middle of our phrase, as its root meaning suggests, is the common element that, by its nature, links the otherwise disparate terms "open" and "infrastructure." The nature of the evolving technology in this domain, we argue, is best served by privatized and competitive provision; but the nature of communications—the commons, the agora, and the public space—is inherently and inevitably public, and, unlike the provision of cardboard boxes or canned peas, requires special attention. The conjuncture of these two domains in communications policy, we assert, requires a new approach.

Table 1.3 outlines four traditional models for regulation in the communications field, arrayed from the most active to the least active role for government. In reviewing that array, it seems appropriate to position OCI near the middle. Most of the world's nations saw fit to provide postal, telecommunications, and broadcasting systems as government monopolies. The United States was one of the few nations with a private

Table 1.3
Five regulatory traditions

Paradigm	Public role	Private role
Public ownership	Ownership and management of monopoly system	Manufacturing only
Common carriage	Heavy regulation of entry, exit, tariff and management procedures	Ownership and management of monopoly system
Open communications infrastructure	Regulation focuses on maintenance of competition and spectrum allocation	Ownership and management of competitive system
Public trustee	Light regulation of procedure, initial regulation of entry	Ownership and management of quasi-competitive system
Laissez-faire	Minimal regulation of business practice	Ownership and management; level of competition not determined

broadcasting system. (Less than 20 percent of the world's nations, most of them in Latin America, had private broadcasting systems until the 1970s.) The United States was virtually alone in choosing a private tele-communications system. As the forces of technology and globalization exert pressure on other nations to move toward private provision, one might imagine that the experience and regulatory tradition of the United States would prove to be a comparative advantage. But this is not neces-sarily true, for it depends on whether the historical traditions of the "Postal, Telegraph, and Telephone" mindset or of the common-carrier and public-trustee mindset will be the most difficult to break through.

Common carriage, as we have noted, puts both federal and local gov-ernment in the role of arbiter of rates, guarantor of universal service, and general overseer of means and methods. It is a delicate and often frus-trating mechanism for both parties—the regulators and the regulated. The system infrastructure is horrendously complex, and regulators are heavily dependent on information provided by the communications com-panies to determine rules and tariffs.

The public-trustee model derives from the American experience in pri-vate broadcasting. The logic dictates that private vendors in limited com-petition can provide better service than can a publicly managed system. Because the spectrum is a public resource, however, in order to maintain their access to it, broadcasters would need to demonstrate their respon-siveness to the public "interest, convenience, and necessity" at regular intervals. For much of the history of American radio and television broadcasting, station managers conducted pro forma audience surveys, filled out forms about news and public affairs programs, ran editorials, and kept technical logs as part of a sham process of oversight. Virtually no stations lost their licenses for failing to serve the public interest. A few had their licenses revoked for illegal behavior and blatant disregard of technical regulations. But the public-trustee tradition does provide a model for content regulation, as regulators keep a close eye out for viola-tions of decency, political access, and right-of-reply regulations. It is a curious mix of legally limited competition, based on the presumption of spectrum scarcity, cyclical bureaucratic attention to content rules of varied vintage, and a codified disregard for the underlying economics. Proponents of deregulation are unlikely to tout the public-trustee concept as the model for the future, but it does offer a useful exemplar of the difficulty of a heterogeneous system in which regulators are charged with monitoring only a narrowly specified set of behaviors of an otherwise unregulated collection of oligopolists.

The laissez-faire model is the ideal to which all policy analysts might otherwise aspire, relying on the marketplace and open competition to hold down prices, spur innovation, and maintain service quality—were it only so easy. The problem of course is the diversity and perversity of market failures—anti-competitive collusion, incomplete market information, distorted transaction costs, and a variety of non-economic social and political goals underresourced by market allocation. These market failures become especially difficult to unravel given the peculiar characteristics of networks: their positive network externalities, the critical importance of interfaces, and their diversity of use.

We focus on the issue of anti-competitive collusion reminiscent of the robber-baron era and put forward a new paradigm of regulation that sets as its goal the maintenance and protection of competition. We identify a series of private mechanisms to deal with incomplete market information and distorting transaction costs. We set aside the issue of other social goals as one best addressed as a matter common to all sectors and not limited to communications.

Perhaps our strongest case is our critique of the halfway evolution of partially regulated communications systems—the worst possible combination of unstructured quasi-random regulation and quasi-unregulated monopolists.

The central elements of the OCI model are:

1. Open architecture.   The future of the network is digital. Voice, text, data, and images will be transported by electrons and photons in digital format. Some messages will require extensive bandwidth, others will not. All communications systems, spectrum- or wireline-based, have the potential for two-way interconnectivity. Some applications such as broadcast-style video may be predominantly one-way, but there is no reason to design the architecture so that two-way communication is precluded (which is true in today's radio, television, and cable delivery systems). At the heart of this approach is flexibility, or the design of a system that facilitates interconnection among different systems and services today and as they develop over time. By emphasizing interconnectivity and interoperability, incentives in innovation shift from trying to protect licensing revenues (by means of a proprietary technical communications standard or inherited spectrum allocation) to encouraging economies of scale in efficient manufacturing and service provision.

2. *Open access.*   The currently dominant model of communications economics is based on a presumption that limited spectrum and inefficiencies of the competitive provision of wireline communications services require legal barriers to competitive entry and regulatory oversight of the legally designated monopolists. This premise, however, is outmoded. The critical turning point occurred first in voice telephony, and will occur soon in radio and in video. Ironically, we see the turning point first in Eastern Europe and the developing world, where the need for building telecommunications is most urgent. Wireless access to the local loop, in many cases, turns out to be fully competitive with wireline provision, but the use of wireless access to the network has other properties, especially in cellular and microcellular applications.

The critical property, we argue, is the lack of steep economies of scale in service provision. To offer universal wireline service from the outset, one must wire every street on every block. To offer cellular service, one can start on a small scale. For fewer customers, fewer cells are required. As service demand grows, the number of cells expands gradually and efficiently, without requiring additional bandwidth. Other properties of course include the mobility and flexibility of personalized communications service. In this way, the promise of Open Communications Infrastructure is in part an unanticipated artifact of cellular and personal communications network (PCN) technology. Wireless access to the local loop makes open competition in provision of local-exchange service a meaningful prospect.

3. *Universal access.*   The principal element of the landmark Kingsbury Commitment of 1913 was a horsetrade. AT&T chief executive officer Theodore Vail was willing to put the entire AT&T network under the control of federal common-carrier regulation in return for protection from further competition. Part of the deal, however, was Vail's commitment to deliver on universal service, with cost averaging; and for the most part, the system he built delivered what was promised.

The deregulation of communications at this juncture raises fears that less privileged and more remote communities will be deprived of access—that competitive provision will lead to the decline of universal service. Critics, for example, point to the deregulation of airlines and the resultant changes in the cost and quantity of service to remote areas. It is at this point, however, that the transportation-communications parallel breaks down. The economies of scale in providing airplanes and airports are still considerable, whereas wireless works the other way. In fact, wireless is much more efficient than wireline for low-density applications in rural areas. Furthermore, direct satellite provision offers additional avenues for service extension and system redundancy. Universal access is therefore a natural component of Open Communications Infrastructure.

4. *Flexible access.*   An interconnected and interoperable network of tele-communications, broadcasting, and electronic publishing breaks down a number of traditional distinctions. In OCI, one can make a call on the cable, send a fax over the radio, and watch television from the telephone line. The format will be digital. The bandwidth will be adjusted according to the demands of the user and the character of the communications. A teenager may not require high-speed and sophisticated data communications to manage a bank account, so the terminal can be simple and inexpensive, and the demand on the network small and economical. Viewers who are casual about the quality of the picture and the release date of the movies they watch on TV can opt for the least expensive display, lower-bandwidth provision, and older movies on an advertising-supported "free" channel or lower-cost pay-per-view. Viewers who are video aficionados with the latest equipment may demand the most sophisticated high-bandwidth signal and recently released programming. Such distinctions need not be legislated: a truly competitive market will provide more diversity than the law could imagine. But the fundamental flexibility and interoperability in communications architecture is not an inevitable outgrowth of current market dynamics. The role of public policy in defining the nature of the market is critical.

The Telecommunications Act of 1996

In the preface we described ourselves as war correspondents, trying to make sense of the industrial battle of the titans as publishers, broadcasters, telcos, computer companies, and new players try to jockey for position, steal their competitors' business, or at least survive with a viable piece of the market they now dominate. They combat each other in the marketplace when permitted, but even more often the fight takes place in the corridors of power in Washington, as they seek to prevent competitors' entry into their traditional markets.

As this book was being written, the battle raged on and ultimately crystallized around a single, massive piece of legislation, Public Law 96-

104, or the Communications Act of 1996. Each week through 1994 and 1995, there would be fresh conflicting predictions that the evolving bill would succeed or become mired in legislative gridlock. But as it turns out, act or no act, policy gridlock continues. In fact, the illusion of a policy of competition and deregulation created by some of the bill's sponsors makes our arguments all the more important.

There is no doubt that the act is a landmark piece of legislation for the twenty-first century. Unfortunately, although the rhetoric is right, the rules are wrong. It regulates more than it deregulates. It delays and constrains competition rather than encouraging it. It is a landmark of micromanagement. By accumulating each powerful lobby's request to tilt the playing field to their relative advantage, it inadvertently introduces a new level of government interference in virtually all forms of electronic communication. It became, in the lingo of the Beltway, a classic "Christmas tree"—a bill to which each industry group attached its own self-serving ornament.

Moreover, activists challenged the provisions for criminalization of indecent content potentially available to children on free-speech grounds within hours of the bill's being signed into law on 8 February 1996. We expect further court challenges to other elements of the act, and additional court proceedings in response to the Commission's mandates, inquiries, and findings.

The act itself is intended to "provide for an orderly transition from regulated markets to competitive and deregulated telecommunications markets consistent with the public interest, convenience, and necessity." In reality that translates into inventing a whole new class of regulations to manage the transition to deregulation. The size of the bill alone tells the real story—over 100 pages of detailed prohibitions and regulations. The term "deregulation" appears twice, the term "regulation" (or a cognate such as "regulatory"), 202 times. There are 353 specific references to the Federal Communications Commission, including 94 cases of "The Commission shall" and 30 cases of "The Commission may." There are 80 formal proceedings that the Commission must initiate.

Yet the publicly pronounced intent of the act is laudable:

To provide for a pro-competitive, deregulatory national policy framework designed to accelerate rapidly private sector deployment of advanced telecommunications and information technologies and services to all Americans by opening all telecommunications markets to competition, and for other purposes (*Congressional Record,* 2/1/1996, p/ H1145).

Furthermore, the rhetorical means of the act are laudable:

• To promote competition in provision of local telephone service.
• To promote competition in provision of long-distance telephone service.
• To promote competition in provision of multichannel television service.

But the act's language establishes complex rules and timetables for setting up additional rules and procedures to guide and regulate the transition to those end states. This is, we argue, attempting to untie the knot one strand at a time. Here, in broad outline, is the act:

Title I   Telecommunications Services

Attempts to set the ground rules of competitive provision of local telephone service and for permitting local telcos to provide long-distance.

Puts forward guidelines requiring existing local carriers to interconnect with potential competitors and unbundle services, details to be specified by the FCC within six months (rural telcos exempted).

Calls for establishment of joint state-federal board to address evolving definition of universal service and creation of special fund to subsidize remote and high-cost service areas.

Calls for the FCC to preempt any state regulations that may inhibit competition.

Sets up detailed checklist of business practices involving interconnection with competitors with which local telcos must demonstrate compliance in order to be permitted by the FCC to provide long-distance service.

Permits local telcos to manufacture equipment through arms-length subsidiaries, dropping last requirement of Modified Final Judgment of the 1982 AT&T antitrust settlement.

Permits local telcos to engage in limited electronic publishing through arms-length subsidiary, but prohibits a long list of possible electronic services such as video, voice storage and retrieval, and video games; these restrictions expire in four years.

Permits public utility companies to provide competitive telecommunications services.

Includes numerous other miscellaneous rules requiring, for example, national geographic rate averaging for long-distance services; special rates for small businesses, disabled parties, and health care and educational institutions; and prohibition of subsidy of pay phones or provision of alarm services.

Title II   Broadcast Services

Authorizes the FCC to grant licenses to existing broadcasters for advanced television services (HDTV) and permits the provision of ancillary (nonbroadcast) services under rules and conditions yet to be determined by the Congress and the Commission.

Requires that for-profit nonbroadcast services incur fees equivalent to that which would have been recovered if the spectrum had been auctioned.

Requires that the spectrum currently used by broadcasters for traditional transmission be surrendered for reallocation under rules and procedures to be determined by the Commission.

Relaxes but does not abolish current media concentration rules by permitting ownership of television stations by a single entity reaching no more than 35 percent of the American viewing audience, and permitting unlimited ownership of radio station groups but limiting ownership in a single market to between five and eight stations (depending on market size). Rules on ownership of multiple television stations in a single market.

Extends broadcast license terms and relaxes renewal procedures.

Title III   Cable Services

Continues authorization of cable rate regulation under new procedures for complaint review and accounting procedures and sets an expiration date for rate regulation in 1999.

Sets up a complex definition of "effective competition" in multichannel television provision and exempts cable providers from rate regulation if the definition is met. Direct broadcast satellite (DBS) multichannel television is explicitly excluded from the definition of effective competition.

Sets up a new category of multichannel television service called "open video systems" similar to previous video-dialtone regulations, which permit common-carriage-like provision with fewer regulations.

Exempts small cable operators from rate and cross-ownership regulations.

Permits limited telco cross-ownership of cable companies and permits telco provision of multichannel television with different regulations if spectrum or wireline facilities are used.

Requires that set-top box technologies be made available from vendors other than cable service providers.

Mandates FCC inquiries into closed captioning and audio television services for the visually impaired.

Title IV   Regulatory Reform

Permits the FCC to waive some provisions of this act if it determines that enforcement is not required to ensure just and reasonable charges for the protection of consumers and the provision of universal service; but the Commission may not waive the local telephone company competitive checklists prescribed under Title I.

Requires the Commission to conduct biennial reviews of the impact of the act's regulations.

Title V   Obscenity and Violence (Communications Decency Act of 1996)

Restricts obscene or harassing use of "telecommunications device."

Requires scrambling of sexually explicit programming on cable.

Creates criminal penalties for anyone who knowingly transmits obscene materials on an interactive computer service.

Permits computer systems operators to block information consumers find objectionable but frees them from liability resulting from possible transmission of objectionable materials.

Requires transmission of a rating code that indicates presence of explicit sexuality and violence in programming and mandates television set manufacturers of sets over 13 inches in diameter to include a VChip to detect codes, thus permitting parental control over children's viewing.

Title VI   Effects on Other Laws

Supersedes the 1982 AT&T Consent Decree and related decrees.

Preserves relevance of antitrust legal tradition.

Removes the FCC from reviewing antitrust ramifications of telco mergers.

Restricts taxation of direct broadcast satellite services.

Title VII   Other

Forbids switching of long-distance unauthorized by customers (slamming).

Addresses regulations on pole attachments and zoning of satellite antennas.

Requires the FCC to facilitate development of advanced telecommunications services especially to educational and medical facilities and public sector institutions.

Sets up Telecommunications Development Fund to promote access for small businesses, rural, and underserved urban areas.

The act consistently changes which rules apply depending on which actor is providing service. The very titles of the act reflect the different rules proposed for telephone, broadcast, cable, computer network, satellite, and public utility providers. Broadcasters (may) get free spectrum for ancillary use (including telecommunications), but telephone companies are expected to bid at auction. The rules about obscene or indecent communications via computer network are different from broadcast, from cable, and from telecommunications services. The VChip provisions apply to television, but not video over the Internet.

The act requires the Commission to make numerous decisions about the existence or absence of effective competition, the fairness of prices, possible discrimination among vendors, possible cross-subsidy between different types of communications services, the technical viability of interoperability, the physical location of telecommunications switching facilities, the character of communications content, and the impact of regulation on market behavior.

Take for example the following language of the act from Title 1, Section 274, which attempts to define what is and is not permissible "electronic publishing" by a Bell Operating Company. The intent presumably is to protect new competitive entrants from the economies of scale and marketing clout of the existing telcos, but the regulatory complexity approaches the absurd:

Electronic Publishing by Bell Operating Companies

(a) Limitations.—No Bell operating company or any affiliate may engage in the provision of electronic publishing that is disseminated by means of such Bell operating company's or any of its affiliates' basic telephone service, except that nothing in this section shall prohibit a separated affiliate or electronic publishing joint venture operated in accordance with this section from engaging in the provision of electronic publishing.

(b) Separated Affiliate or Electronic Publishing Joint Venture Requirements.—A separated affiliate or electronic publishing joint venture shall be operated independently from the Bell operating company. Such separated affiliate or joint venture and the Bell operating company to which it is affiliated shall:

(1) maintain separate books, records, and accounts and prepare separate financial statements;

(2) not incur debt in a manner that would permit a creditor of the separated affiliate or joint venture upon default to have recourse to the assets of the Bell operating company;

(3) carry out transactions (A) in a manner consistent with such independence, (B) pursuant to written contracts or tariffs that are filed with the Commission and made publicly available, and (C) in a manner that is auditable in accordance with generally accepted auditing standards;

(4) value any assets that are transferred directly or indirectly from the Bell operating company to a separated affiliate or joint venture, and record any transactions by which such assets are transferred, in accordance with such regulations as may be prescribed by the Commission or a State commission to prevent improper cross subsidiaries;

(5) between a separated affiliate and a Bell operating company—(A) have no officers, directors, and employees in common after the effective date of this section; and (B) own no property in common;

(6) not use for the marketing of any product or service of the separated affiliate or joint venture, the name, trademarks, or service marks of an existing Bell operating company except for names, trademarks, or service marks that are owned by the entity that owns or controls the Bell operating company;

(7) not permit the Bell operating company—(A) to perform hiring or training of personnel on behalf of a separated affiliate; (B) to perform the purchasing, installation, or maintenance of equipment on behalf of a separated affiliate, except for telephone service that it provides under tariff or contract subject to the provisions of this section; or (C) to perform research and development on behalf of a separated affiliate;

(8) each have performed annually a compliance review—(A) that is conducted by an independent entity for the purpose of determining compliance during the preceding calendar year with any provision of this section; and (B) the results of which are maintained by the separated affiliate or joint venture and the Bell operating company for a period of 5 years subject to review by any lawful authority; and

(9) within 90 days of receiving a review described in paragraph (8), file a report of any exceptions and corrective action with the Commission and allow any person to inspect and copy such report subject to reasonable safeguards to protect any proprietary information contained in such report from being used for purposes other than to enforce or pursue remedies under this section.

(c) Joint Marketing.—

(1) In general.–Except as provided in paragraph (2)—(A) a Bell operating company shall not carry out any promotion, marketing, sales, or advertising for or in conjunction with a separated affiliate; and (B) a Bell operating company shall not carry out any promotion, marketing, sales, or advertising for or in conjunction with an affiliate that is related to the provision of electronic publishing.

(2) Permissible joint activities.—(A) Joint telemarketing.—A Bell operating company may provide inbound telemarketing or referral services related to the provision of electronic publishing for a separated affiliate, electronic publishing joint venture, affiliate, or unaffiliated electronic publisher, provided that if such services are provided to a separated affiliate, electronic publishing joint venture, or affiliate, such services shall be made available to all electronic publishers on request, on nondiscriminatory terms. (B) Teaming arrangements.—A Bell operating company may engage in nondiscriminatory teaming or business arrangements to engage in electronic publishing with any separated affiliate or with any other electronic publisher if (i) the Bell operating company only provides facilities, services, and basic telephone service information as authorized by this section, and (ii) the Bell operating company does not own such teaming or business arrangement. (C) Electronic publishing joint ventures.—A Bell operating company or affiliate may participate on a non-exclusive basis in electronic publishing joint ventures with entities that are not a Bell operating company, affiliate, or separated affiliate to provide electronic publishing services, if the Bell operating company or affiliate has not more than a 50 percent direct or indirect equity interest (or the equivalent thereof) or the right to more than 50 percent of the gross revenues under a revenue sharing or royalty agreement in any electronic publishing joint venture. Officers and employees of a Bell operating company or affiliate participating in an electronic publishing joint venture may not have more than 50 percent of the voting control over the electronic publishing joint venture. In the case of joint ventures with small, local electronic publishers, the Commission for good cause shown may authorize the Bell operating company or affiliate to have a larger equity interest, revenue share, or voting control but not to exceed 80 percent. A Bell operating company participating in an electronic publishing joint venture may provide promotion, marketing, sales, or advertising personnel and services to such joint venture. (D) Bell Operating Company Requirement.—A Bell operating company under common ownership or control with a separated affiliate or electronic publishing joint venture shall provide network access and interconnections for basic telephone service to electronic publishers at just and reasonable rates that

are tariffed (so long as rates for such services are subject to regulation) and that are not higher on a per-unit basis than those charged for such services to any other electronic publisher or any separated affiliate engaged in electronic publishing.

Or review the following language, which attempts to make it very clear how rural telecommunications service providers are to be treated with special care:

(1) Exemption for certain rural telephone companies.—(A) Exemption.—Subsection (c) of this section shall not apply to a rural telephone company until (i) such company has received a bona fide request for interconnection, services, or network elements, and (ii) the State commission determines (under subparagraph (B)) that such request is not unduly economically burdensome, is technically feasible, and is consistent with section 254 (other than subsections (b)(7) and (c)(1)(D) thereof).—(B) State termination of exemption and implementation schedule.—The party making a bona fide request of a rural telephone company for interconnection, services, or network elements shall submit a notice of its request to the State commission. The State commission shall conduct an inquiry for the purpose of determining whether to terminate the exemption under subparagraph (A). Within 120 days after the State commission receives notice of the request, the State commission shall terminate the exemption if the request is not unduly economically burdensome, is technically feasible, and is consistent with section 254 (other than subsections (b)(7) and (c)(1)(D) thereof). Upon termination of the exemption, a State commission shall establish an implementation schedule for compliance with the request that is consistent in time and manner with Commission regulations.

Time and again, the act substitutes political muscle for public interest in determining policy. Take, for example, the definition of effective competition for deregulation of cable television. Multichannel television competition from a telephone company counts as effective competition, but the same service provided by a satellite service does not. Why? Well, if one counted competition from satellite broadcasting in the United States in 1996, virtually all markets would qualify as effectively competitive because of the existence of DirectTV and PrimeStar multichannel direct broadcast satellite service offerings.

Similarly, the act introduces the idea of competition in local telecommunications with great grandeur, but local competition was already under way in a dozen states under a variety of local deregulatory experiments.

Will the public interest in universal service be adequately protected in the act? A complicated fund for providing competitive service for regions not likely to get competition (and hence, lower rates) is embodied. But just who benefits from this—and what it does for those left out of the information revolution due to lack of money, education, or interest—is quite unclear. Assumptions that competition will lower costs without pushing margins to the razor's edge were clearly behind the grand view of the bills of the past several years, but a more compelling problem has arisen in the meantime: how does either the industry or the government finance future telecom infrastructure if carriage approaches zero cost as capacity approaches infinity, and if markets for carrying bits contract because of new technologies rather than expand. This is not conventional wisdom for conventional economics; the act may be much too late.

The questions at the moment are: (1) whether the deals struck will remain acceptable to a confused Congress and a perplexed administration during and after a particularly strident election year; and (2) whether the large long-distance firms (AT&T, MCI, Sprint) have sufficient market presence to withstand the assault from equally powerful new competitors without the albatross of obsolete plant as they enter the competitors' local market, which has a similar set of technical burdens. It is likely that a new, though perhaps not as comprehensive, bill may be crafted in the near future. If new deals cannot be struck among these companies, different arrangements will be proposed by a future administration and a different Congress. As local rates zoom where traffic remains stagnant, all sorts of unexpected anomalies may start driving new entrants and the general public to distraction. The sweet smell of success may turn sour quite rapidly unless some miracle creating new markets, new jobs, and new money appears rather quickly.

The Clinton administration, in spite of its unhappiness with provisions on cable television rate deregulation and purportedly with matters relating to universal service and attacks on freedom of speech, needed a bill to demonstrate its continued leadership in national and Global Information Infrastructure development. It felt it had to confirm its ability to work effectively with the Republican leadership of the House and Senate on matters important to the business community. The long-distance carriers, recognizing that they had been outmaneuvered by the local telephone firms in gaining the confidence and support of key Republican senators and congresspersons, chose finally not to fight a battle that might lead to greater estrangement with the Republican leadership. Externalities carried the day; change was inevitable, and the mood was that either a bill passed then or never. With the potential markets and technology curves equally confusing to all, whatever entrepreneurial spirit was emerging in a formerly wholly monopolistic, quite non-market-oriented set of industries took over. Working in unusual secrecy with the House-Senate conference committee (which finally limits lobbying opportunities to only the most connected players), they reconciled on a bill that even Clinton could not find enough political strength to veto, despite some clauses that were anathema to the Democratic party's long history.

The principal failure of the act is the clear lack of understanding shown regarding the real progress being made toward development of the National Information Infrastructure through the incredibly rapid expansion of internetworked small and distributed computer processing—the Internet. The Internet's main feature—a "bug" to those who like things orderly and predictable—is that the net is uncontrollable, even chaotic. Its novel and radical applications, evolving open protocols, growth of users and traffic, and most important, its unpredictable change, has become a global driving force for information infrastructure. It is both scary (as the Chinese government has demonstrated by its regressive laws) and exhilarating (as the stock market has indicated by multiples in the stratosphere for certain companies with dubious products and negative income flows).

For example, on the Internet:

• traffic can shift suddenly;
• services can change suddenly;
• new technology and potential de facto standards can be introduced by anyone, as long as they work;
• the only effective "governing body" is the all-volunteer group of experts called the Internet Engineering Task Force, which deals with technical protocol design and maintenance;

• a residual coordinating power remains with the U.S. government through R&D programs of the Department of Defense Advanced Research Projects Agency and the National Science Foundation—but both of these agencies are by their nature averse to regulatory policymaking. Due to pressure from the formerly dominant computer and telecom firms, the Federal government is easing out of one area after another related to the "governance" of the Internet. At the same time, the Federal Communications Commission and the Federal Trade Commission are increasingly questioning how their traditional regulatory activities are being affected by the growth of the Internet. Instead of direction being taken up by the old-line conventional firms, the past few years have shown even more chaos and radically innovative ideas coming from small firms (that become big firms overnight) such as Netscape, Newbridge, Sun, Cisco, Intel, Cascade, and Microsoft. A decade ago, who would have predicted that such firms would dominate with power and ideas? What have AT&T, IBM, and DEC contributed that shifted the path of information infrastructure in comparison with what the new giants on the block have done? From one cynical point of view, it can be said that the old firms contributed the laid-off personnel that helped direct the new leading companies.

The lack of understanding in Congress and among the special-interest lobbying groups may be the biggest surprise of the early years of the next millennium. The delay and equivocation that will certainly emerge from lawsuits arising out of the act's contradictory provisions cannot be avoided in the U.S. political process. While older vested interests (and the jobs they encompass) may realize they bought a devil's bargain, telecommunications lawyers and consultants are not threatened by the legislation in any way and can look forward to a prosperous and fruitful future under any scenario.

The Law of Unintended Consequences

Another likely effect of the act is that it will have significant recursive effects on technology, market development, and policy practices: the law of unintended consequences. In this, the recursive nature of digital computers themselves—that a stored-program device, as Alan Turing understood from the very beginning some sixty years ago, can define its own instructions, its own cyberspace, so to speak—are a natural outgrowth of a chaotic system under no one's control.

It is unlikely that the technical, economic, and policy distortions introduced by the Telecommunications Act of 1996 will inhibit the rapid development of the National and Global Information Infrastructures, but the realization of universal service and its social benefits for education and health care, to name just two areas, are unlikely to be either equitable or substantive without some major modifications in the near future. It certainly was not the intention of the bill's drafters to spread the new wealth, but the public has yet to be heard. As with other concerted public movements in the past—ranging from the original pressures to regulate and rein in the railroads and telecom firms in the past century, to antimonopoly actions in this century, to the current powerful and increasingly popular grass-roots movement for environmental protection and change—telecommunications is sure to be affected by the reaction of its customers to the new provisions. Unlike some other industries that labor and profit in quiet sectors, communications deals with everyone, all the time, and in insidious ways. The actors cannot hide.

But aside from the potential unintended consequences of the act, there are the obvious embodied flaws: Parts of the act include unconstitutional, and to some unconscionable, abridgments of the freedom of speech, the press, religion, and the right to petition the government for a redress of grievances. The history of the laws protecting speech (and it has been noted that the First Amendment is but a local ordinance in cyberspace) indicates that courts in all democratic countries take a narrow view of government interference in such rights because through bitter experience, democratic citizenry do not trust their governments. Such lack of trust extends to the denizens of government bureaucracies themselves, who know how greatly power can corrupt. Arguments about speech will not simply go away because of ill-conceived legislation. Indeed, since the act was passed the level of "obscene" speech on the Net has probably increased as "netizens" feel compelled to challenge the act's restrictions.

One unintended consequence of the act is that the legal separation and traditional treatment of the printed press, broadcasting, and telecommunications may have collapsed in a new framework that pretends to provide a consistent treatment for all media. This was the key point Ithiel de Sola Pool made in his landmark study, *Technologies of Freedom* (Pool 1983). As we noted, how does one categorize "multicasting" under the new law? Another unintended result: if the separate treatment of broadcasting and telecommunications no longer stands up to legal scrutiny, and if government intrusion into private communications between individuals and groups is to be sanctioned, then the effects of the act may be either incredibly repressive, or if thrown out by the courts, a useless exercise.

In the Internet community, it is said that "the Internet treats censorship as a broken connection and works around it." It is important to understand that this is more than an allusion—the Internet Protocol (IP) does exactly that, and to program servers extending the route with Web software is simple—even a grade-school student can learn how. Perhaps another unintended consequence of the act is to raise the status of technical prodigies to political heroes. Who said government doesn't know how to instill educational goals and create new jobs?

The futile attempt in the Congress to pander to an ill-informed public about the real, but far from pervasive or intrusive, dangers of on-line pornography is one pertinent example of how far congressional and press attention was removed from the critical question of whether the cross-sectoral deals struck in the act will really serve to "provide for an orderly transition from regulated markets to competitive and deregulated telecommunications markets consistent with the public interest, convenience, and necessity."

More appropriate than the provisions of the act for governing the requirements of an information-intensive society would be a new policy framework for an Open Communications Infrastructure. This new framework would make no distinction between wired and wireless networks or between content and conduit. Only such a unified policy framework is capable of supporting social and economic needs as well as sustaining technical development in the years to come. This act falls far short of this goal.

Conclusion

It is a perilous venture to write about communications policy when politicians, lobbyists, and the daily papers assure us that the key issues have been resolved by the new act. We think not. For, as we have argued, neither the myopic advocates of incremental changes to the status quo nor the breathless futurists and cyberspace libertarians have adequate understanding of Open Communications Infrastructure. One side would have us inch forward (or backward) on the information highway, the other would have us wander, lost in cyberspace without a map or any shared understanding of the rules of the road. In the following chapters we review the basis in technology, economics, and policy for our position.

2

The Nature of Networks

Perhaps the information superhighway rubric has become so popular because it resonates with people's idealized sense of what a network ought to be: Access is easy, and coverage and speed are maximized. Highways, supplanting railroads, canals, and post roads, became the commons of the industrial age. Although built and often maintained by private contractors, highways, like public communications systems, are essentially public entities, built with public resources for the public good. Furthermore, the hub-and-spoke architecture of transportation systems was the model for telegraphy, telephony, and broadcasting. Although historically apt, the highway metaphor could prove to be a misleading model for public communications policy, because modern digital communications systems do not work like physical highways. Their architectures, load dynamics, and economic properties are unique. We will explore the nature of networks—when the highway and infrastructure metaphors are useful for understanding the communications revolution, and when they are not.

Roads and Railways
The railroads, because they required different physical links than those of any previous infrastructure, were the first of several industrial innovations to exploit and redefine networks and networking. As we discuss in chapter 3, railways evolved from little more than wooden guide rails laid in the public highway, and as Schivelbusch described, "as the motion of transportation was freed from its organic fetters by steam power, its relationship to the space it covered changed quite radically" (Schivelbusch 1986).

Steam defined a break with the past, and as we argue later, digital telecom bits signal a new break with older analog telecom systems. And, as with analog devices, we may perceive that "preindustrial traffic [was] mimetic of natural phenomena. Ships drifted with water and wind currents, overland motion followed the natural irregularities of the landscape and was determined by physical power of draught animals." So, Schivelbusch demonstrates, "the railroad was not seen as an autonomous traffic system, but as a highway provided with rails which essentially retained its traditional traffic pattern.... The vehicles remained individual road vehicles, but for the duration of their travel on the rails they had to be lifted onto an undercarriage that is fitted to the rails" (Schivelbusch 1986).

As with many sea changes, the purpose of iron rails was seemingly innocent—to reduce friction; it took a leap of faith to imagine that such a little change was to have enormous import on underlying economic structures beyond reducing costs of transport. Thus we can appreciate Adam Smith's complaint that the "upkeep of a horse was equal to the feeding of eight laborers," so "when the one million horses kept for the purposes of transportation in England were made redundant by mechanization, they would release foodstuffs for eight million laborers," begging the question of oats and barley (Smith 1776).

But like the digitization of networks, whose purpose is to make amplification more efficient and transmission more accurate, little changes may usher in revolutions. To operate the new railway networks, new forms of organization and administration were required, finance had to generate volumes of capital beyond what any single bank or government tax-collection agency could raise, and new technologies had to be invented for communication, manufacturing, inventory, and above all else, control. Schivelbusch notes that "in order to guarantee the proper functioning of this machine, juridical and politico-economic regulations had to be revised, but technological improvements proved equally necessary. The most important technological addition to the railways was the electrically operated telegraph system" (Schivelbusch 1986).

At the opening of the Boston-to-Albany railway in 1842—the first to cross the Appalachian chain from the Atlantic seaboard to the hinterland—Governor Everett said, "Let us contemplate the entire railroad, with its cars and engines, as one vast machine! What a portent of art! Its fixed portion a hundred miles long; its movable portion flying across the State like a weaver's shuttle" (Buckingham 1842). The railway, however, proved to be neither a gigantic yet elegant machine nor a simple extrapolation of the highway; network externalities showed up early. By 1840, England's Parliament, for reasons of safety and efficiency, decided that "railway companies using locomotive power possessed a peculiar monopoly for the conveyance of passengers, from the nature of their business" (Jackman 1916). But monopoly status on their own rails was not enough; railway systems still "remained merely the sum of numerous local and regional lines that operated independently from each other ... [and] even working *against* each other" (Schivelbusch 1986).

With the telegraph, the railways could effectively run longer and longer links. Therefore consolidation became technically feasible, and as we know, what is possible often becomes mandatory: the bankers insisted on mergers to protect their investments for a variety of reasons (Adler 1970). So quite rapidly, consolidation and monopoly concepts were extended from the track to the market, de facto if not de jure (Dagget 1908; Johnson 1926; Reed 1969; Schivelbusch 1986).

Control paradigms, whereby control of railway operations implied control of transport markets, was the essence of the nature of the new transport networks (Beniger 1986a). And market control led to market abuses—at least as we would define them today.

Nostalgia for "empire building" aside, the robber barons were not terribly friendly to their environment, their customers, their employees, or their contemporaries. Greed and personal wealth were the motivating factors, not the buildup of infrastructure for the benefit of the country or the creation of new enterprises (Allen 1935; Gordon 1988; Grodinsky 1957; Holbrook 1947; Summers 1993). Such rationalizations came after the fact (along with some measure of philanthropy), and the real personal contributions of the robber barons to the development of the American industrial empire are arguable to this day, considering the hefty monetary and legal subsidies they received from both the state and ordinary taxpayers (Goodrich 1960; Mercer 1982; Rae 1979).

Greed and Gateways
Greed could be furthered most by control of gateway routes, gateway processes, and gateway agencies. The war—and indeed it was a war,

fought with guns and settled only after much bloodshed—between the Santa Fe and Rio Grande railways for control of a key mountain pass in the Rockies exemplified the lengths to which the railroad magnates would go to further their personal interests. This was during a period after the Civil War when numerous conflicts between the empires and emperors were fashioning the early industrial conglomerates. Not all were sorted out by pitched battle; some were settled in the courts, and some by bankrupting rivals. For the railroads, most of the battles were to control some bridge line or another, or to prevent a competitor from completing a parallel line, usually built for blackmail purposes. Change the dates and the weapons, and it sounds much like today's battles among the telecom giants (Beebe 1952; Cummings 1937; Grodinsky 1957; Grodinsky 1962; Quiett 1934).

The Santa Fe-Rio Grande war, which lasted from 1878 to 1879, was illustrative of how things could get out of hand in the absence of established rules and law enforcement. Indeed it was a true American Western saga.

There were only a handful of passes along the eastern slope of the Rockies that could be readily bridged with the technology of the mid-nineteenth century, which included black powder, backbreaking labor, and engines that could negotiate relatively easy grades. Most of these routes had been mapped out by the federal government in a series of expensive and somewhat dangerous survey expeditions by the U.S. Army between 1830 and 1850, all wholly funded by the taxpayer.

But the government did nothing about detailed planning, much less about defining the gateways to the West. They left that up to whoever claimed land first and made their financial deals with the government— an early form of benign neglect. There were several canyons and passes that looked promising, and it was not clear which were best suited for future development. So the diligent railway entrepreneur attempted to stake out all that he could, as we will see, often with guns and blasting powder. The federal government had recommended five general corridors, somewhat equally spaced from north to south, for crossing the Rockies. Eventually four of the routes were completed, each with several lines.

The Denver and Rio Grande Railroad was initially formed to build south to Mexico. The company hired William J. Palmer, a famous Civil

War general and civil engineer, who had built the Pennsylvania Railroad before the war, for the task. Palmer's engineering skills were at least as important in this environment as his military prowess. Like its rival, the Atchison, Topeka, and Santa Fe, the Rio Grande had missed the early window for transcontinental land grants. So Palmer instead planned to build over Raton Pass along the New Mexico boundary, and on the way south would figure out how to turn westerly–national planning by the seat of his saddle, so to speak.

However, "when the Denver and Rio Grande forces arrived [at Raton], the Santa Fe was in armed possession, thus effectively shutting off Palmer's entry into southern territory. After this *coup de main*.... Palmer resolved that any further advantage to be obtained by the Santa Fe would not be without a struggle," one which, as a good general, he knew he had not the resources to win (Quiett 1934).

It was as a result of this critical bottleneck that the railroad war between the Santa Fe and the Rio Grande was fought in earnest at the entrance of the Royal Gorge. The Rio Grande had another advantage beyond its famous general-engineer, because "having possession of the telegraph lines, [Palmer] discovered that the Santa Fe was about to make a sudden dash into the canyon, and a spirited scramble for priority ensued" (Quiett 1934). Small skirmishes occurred over several months during which the rival roads were able to build stone forts at key points and hired gunslingers and unemployed Civil War veterans to man them. But the Santa Fe managed to get into the canyon and begin construction of a bracket bridge to span the roaring creek—despite the Rio Grande's possession of the telegraph lines.

Quiett continues the story:

Armed with writs to sheriffs in every county on the line, Palmer now began a systematic campaign of seizure of the property, as coolly and calculatingly as if he were still "the best cavalry officer in the Army, bar none." A group of armed men was mobilized in East Denver, marched to the general offices [of the Santa Fe], broke open the doors with a tie for a battering-ram, and occupied the place. A passenger train then made up by the new management triumphantly proceeded southward, systematically capturing stations all along the line and taking the captive station-agents aboard.... At Cucharas two Santa Fe men were reported killed and two wounded. At Pueblo the Santa Fe had imported Bat Masterson, famous sheriff of Dodge City, the toughest town in Kansas, to defend its property. This crack-shot sheriff had recruited a band of fighting-men and posted them

(a)

Figure 2.1
Standards battles, nineteenth-century style. One hundred years ago, when railroad magnates could not settle differences of strategy, or track gauge, they
brought out their private armies. (a) Bat Masterson, later to become famous (or
infamous) Federal Marshall, leads a fatal shootout between a small armed band
of Atchison, Topeka & Santa Fe Railroad workers against a similar band of
trackmen employed by the Rio Grande Railroad at the Royal Gorge in Colorado.
(b) The ultimate settlement, dictated by the U.S. Supreme Court, was a dual-gauge,
single track for both lines through this critical pass at the foothills of the Rockies.
(a) Courtesy, Colorado Historical Society (F1599). (b) Courtesy, The Library of
Congress.

(b)

Figure 2.1 (continued)

in the railroad roundhouse ready to pick off the Palmer men as fast as they appeared. But the practical and suave treasurer of the Rio Grande, Robert F. Weitbrec, was aboard the train; reasoning that hired assassins could become peacemakers if offered a higher wage, he waved a flag of truce and succeeded in negotiating a cessation of hostilities with the truculent Bat. (Quiett 1934)

In the terms of modern-day capitalism, it was a successful hostile take-over. (Jensen 1974)

Eventually the courts ruled that the Rio Grande had to return the seized property, and that both railroads had to share the pass. Chronicler

Glenn Quiett concludes, "Thus ended the most exciting episode in Western railroad building" (Quiett 1934). The stage, however, was now set for the new networking paradigm of gateway control, although, we hope with more litigation and less bloodshed than in the Wild West.

Bells and Whistles

The origins of American Telephone and Telegraph (AT&T) in the mid-1880s must be set against this background. Its beginnings are more obscure than most observers have thought (Stehman 1925). The formation of what became the holding company for the largest group of private telephone carriers in the world had less to do with forming a telephone monopoly than with building railroads—not an obvious connection, but one that explains where the "telegraph" appendage came from (Solomon 1978). AT&T never delivered a telegram in its long history, but it did operate a large percentage of private telegraph circuits in the United States, very likely equal to those run by Western Union (although the statistics were not terribly accurate in the early days) (Rhoads 1924).

The fledgling Bell System was only a collection of local telephone franchises, licensed to use the Bell patents and lease phones owned by the American Telephone Company in the first few years after Alexander Graham Bell had filed his controversial patents in 1876 (Casson 1910; Cummings 1937; Danielian 1939; Faulhaber 1987; Rhodes 1929). A number of parties contested their validity, including the dominant communications carrier, Western Union, which by the mid-1880s was controlled by railroad magnate Jay Gould. Gould epitomized the robber baron, right down to his overt bribery of judges and political hacks, and financial shenanigans that today would land him in jail, but at the time only annoyed his rivals. Gould managed to trick the Vanderbilts into trading Western Union in the early 1880s for a critical bridge railroad they needed to strengthen their New York Central and Hudson River Railroad (Grodinsky 1957; O'Connor 1962).

Gould had an interesting tactic in his railroad trading sleight of hand: he would read his business rivals' telegrams. An entire floor of the block-long Western Union complex in lower Manhattan was dedicated to filtering telegrams for juicy tidbits, stock tips, and other items for blackmail or leverage. This was all perfectly legal; indeed, your implied contract when you sent a telegram was that the telegraph company could read your messages to "ensure proper delivery." Gould's judges helped maintain the law on his side, in case there was a dispute (Grodinsky 1957; Oslin 1992).

Gould annoyed his rivals so much that the wealthiest families, among them the Vanderbilts and the Forbeses, finally had to band together against him (using methods that today would be considered criminal conspiracies in violation of antitrust laws) to control his manipulation of the securities markets. What brought things to a head was a major crisis and a minor ploy that no one seemed to notice until it was too late. The seemingly irrational ploy was Gould's method of buying smaller railroads that had important connections, and then shortly thereafter selling them to different interest groups, destabilizing any attempt to create reasonably competitive groups of roads. Competition for traffic was fierce, and each new combine merely split the traffic more and more without increasing the size of the pie. Rates were dropped below marginal cost, so no one was making money. The railroads attempted to cartelize the industry into "pools"—legal then before the Interstate Commerce acts and antitrust laws were passed. But as soon as a pool stabilized a region of the country, Gould would break any existing agreements and rearrange the systems (Grodinsky 1957; Grodinsky 1962; Thompson 1947; MacAvoy 1965).

But Gould turned out to be less interested in railroads than it seemed to his preoccupied rivals. What he was doing was buying roads for their telegraph contracts with competing carriers, turning the contracts over exclusively to Western Union with extremely long, unbreakable leases, and then selling the roads. Pretty soon Western Union became a virtual monopoly. Furthermore, it was able to prevent the nascent Bell Telephone companies from interconnecting via railroad rights-of-way. (Carriage roads before the advent of the automobile were pretty useless for interurban travel or wires.) On top of this, Bell and Western Union were contesting each other's telephone patents in court; and to complicate matters even more, the Bell companies had filed a patent that would allow them to send telegraph signals over telephone lines, should they ever acquire enough intercity circuits to make it worthwhile (Brock 1981; Danielian 1939; Faulhaber 1987; Solomon 1978).

With the railroad wars about routes and below-cost rates, the extreme complexity of telephone patents (which were finally settled in Bell's interest in 1888 in the longest Supreme Court case ever), and Gould's manipulation of the securities markets with his access to Western Union's telegrams, a form of gridlock developed among the most powerful infrastructure interests in the United States. This is similar to today's diversions over regulation, pornography, and spectrum, which contribute to the political gridlock of our epoch.

Enter young J. P. Morgan, scion of an English banking family, who was about to make his name in America with his father's seed money and U.S. business connections. In 1885, Morgan invited all of Gould's enemies, who happened to be feuding with each other, to his yacht, *Corsair,* anchored in Long Island Sound. No one was allowed to leave until they settled the railroad/telegraph/telephone wars (a nineteenth-century example of today's digital convergence). The railroads agreed to cartelize a stable-route oligopoly; to keep rates high enough for everyone to make a profit; to stop raiding each other's territories; and in a famous trade, Vanderbilt's New York Central abandoned a partially constructed right-of-way (route of today's Pennsylvania Turnpike) parallel to its main rival, the Pennsylvania Railroad, in exchange for the Pennsylvania Railroad selling the West Shore Railroad on the opposite side of the Hudson and Mohawk Rivers to the New York Central (Beebe 1952; Holbrook 1947; Sinclair 1981; Winkler 1930).

One of the less noticed results at the time was the formation of AT&T, under the control of the Forbes family, to counterbalance Gould's rapacious Western Union. Gould gave up Western Union's telephone patent claims in exchange for the new telephone's promise to keep out of telegrams (but that did not stop the Forbeses from becoming significant players in private-line telegraphy) (Cummings 1937; Danielian 1939; Grodinsky 1957; Solomon 1978; Pier 1953).

Less than two years later, however, the railroad deals of 1885 fell apart, and Congress came under increasing pressure to form the Interstate Commerce Commission to bring some semblance of civility and predictability to the marketplace (Horwitz 1989; Kirkland 1965; Kolko 1963; Locklin 1954). Transportation infrastructure was too important for the robber barons to control. The telecom cartel held out a bit longer. But the Forbes' Bell Telephone patents had expired by 1892, and rival companies began to form again, including some using telegraph rights-of-way to parallel AT&T's wholly owned Long Lines subsidiary (Danielian 1939). The new telegraph/telephone and railroad wars were in full swing by the turn of the century, and new federal and state regulatory and antitrust efforts had to be brought to bear to calm down and recartelize infrastructure—this time under the cloak of legality (Horwitz 1989). For AT&T, regulation came in the form of the 1913 Kingsbury Commitment, which settled monopoly rights until the 1980s.

This pattern of deal making, and deal failure, repeated itself every quarter century or so, each time yielding greater and tighter governmental regulation than before; the deals seldom held, and the greedy, it would appear, just became more so.

These are recurring cycles of new deals designed to break gridlock: manic mergers between telcoms and cable companies, secret efforts to reinstate natural monopoly (at least on a local level), intellectual property rights claims for software that does not even exist yet. In other words, pressures for deals occur wherever technology casts a dark shadow, making the future hard to see, just like the creation of the American Telephone and *Telegraph* Company in 1885, the Kingsbury Commitment of 1913, the Radio Corporation of America cabal of 1926, the Federal Communications Act of 1934, the AT&T antitrust "pre-judgment" of 1949, the so-called Final Judgment of 1956, and the *Modified* Final Judgment of 1982.

Nature and Networks

A network is more than an assemblage of randomly connected links and junctions; how they connect is important. Connectivity is the difference between a truly national (and international) information infrastructure and today's balkanized structure of communications and information systems. That is part of the reason why the highway analogy is so seductive; virtually every road is connected to every other road, vehicle sizes are standardized so that practically any vehicle can utilize the system, and financing and maintaining the system has become routinized and relatively noncontroversial. Few entrepreneurs perceive toll roads to be the

(a)

Figure 2.2
The dealmakers and their craft. When magnates cannot agree and bullets do not work, the corporate leaders look for one of their own to make a deal. (a) In the late 1880s, J. P. Morgan, the young American-born scion of a British banking family, maintained friendships and financial connections with a host of rival railroad executives. (b) During a particularly intense price war, Morgan invited the key players to his yacht, the Corsair, in Long Island Sound, and reportedly would not let them off until they cut a deal. The deal did not hold and in the next decades Congress and the courts provided new rules of engagement. A century later, in 1993, cable television CEO John Malone of TCI invited Ray Smith, CEO of Bell Atlantic, to his yacht off the coast of Maine to discuss a possible mega-merger of their companies into a multimedia giant. The deal was made and publicly announced with great fanfare, but fell through later that year. (a) Culver Pictures.

(b)

Figure 2.2 (continued)

economic gold mine of the future, so there is little hubbub over privatization (although highway privatization is now being pursued in, for example, Mexico and Argentina). We take it for granted that roads are public and vehicles are private. It is, as they say, a done deal. In the evolving electronic network, however, all bets are off, for a variety of reasons:

• Connecting to the network is not standardized.
• Different industrial sectors promote different and noninteroperable architectures, and many find it in their interest to prevent or delay rather than promote interconnectivity.
• Service users and service vendors are locked in a battle for control of network parameters and economics.
• The role of private and public entities in the design and management of the system is highly controversial.

• Many of the players see these technologies and services as the economic gold mine of the next several decades and strategize accordingly.

A key distinction can be made between physical networks and logical networks. Most of the generic properties of physical networks were shared by both the transportation and communications systems until the 1980s (Solomon 1989d; Solomon 1990a). The application of computer power to communications—the widely touted digital revolution—requires a new paradigm to understand the technical architecture, economic properties, and appropriate policies for an entirely new type of network. This is a classic case of a paradigm shift, as the technology forces a reconceptualization of how business is conducted (Solomon 1991). The underlying economics and appropriate policy for regulating physical networks have become routinized and are relatively well understood. At the close of the twentieth century, however, we are witnessing a period of chaotic transition as inventors, entrepreneurs, captains of industry, and government officials hammer out a new set of standard operating procedures for conducting business on the electronic commons. We can learn a great deal from economic history, but we must not become captives of its received wisdom and routinized assumptions and thus fail to recognize the unique characteristics of electronic communications infrastructure.

Transportation Networks: The Evolution of Hierarchical Architectures

The major industrial transition of our time is the shift from moving objects over waterways, bridges, rails, roads, and airways to moving symbols over electronic networks. Although one could argue that this progression has been under way since the beginning of the industrial revolution (Beniger 1986a), it has been accelerating in pace and significance during the past two decades in the wake of radical changes in underlying communications and related computing technologies (Dordick 1981; Neuman 1991).

The model for infrastructure in the industrializing nineteenth century was to improve the agora, the central physical marketplace. The key architectural principles were centralization and hierarchy, moving from the hub, that is, a central regional trading center such as Chicago or St.

Louis, down radiating spokes to subsidiary marketplaces of decreasing size. The canals and then the railways grew up around the existing trading centers on rivers and ocean ports. Industrialists built centralized transport depots, grain elevators, manufacturing facilities, and supporting financial and administrative institutions. That is, of course, where the first telegraph offices and later the central telephone switching offices were located.

One reason that transport models guided telecommunications architecture is that the railroads urgently needed telegraphic communication to avoid accidents. The efficient use of a single railroad track for traffic in both directions over long distances was not possible with physical line-of-sight signaling. Several dramatic head-on train collisions in the 1840s demonstrated the need to put this new telegraphy invention to work—so for much of this era the telegraph lines literally ran along the tracks (Beniger 1986a; Coe 1993; Rhoads 1924; Thompson 1947).

But perhaps the major reason that transportation and communications architectures were so similar is that they had common economic and physical properties. A review of the generic properties of nineteenth- and early twentieth-century transportation networks illustrates several interesting features:

• the speed of transport was generally quite slow;
• transportation networks had highly constrained capacities;
• networks had high capital costs for construction, engineering, and maintenance, so the cost of transport remained a relatively high proportion of the cost of goods;
• transportation costs were highly sensitive to distance;
• construction costs were highly sensitive to geography;
• networks tended to be inflexible and unchanging due to large sunk costs;
• networks were often unreliable;
• networks were generally inefficient;
• parallel, noninteroperable, specialized carriers for special purposes (such as coal barges) were economically viable due to low efficiencies of the public rail network.

As a result of these generic properties, economies of scale, scope, and market power reinforced incentives for centralized ownership, control, and systems architecture, as well as the regulation of these large systems by variations on the common law tradition of common carriage (Horwitz 1989).

By World War I, the pattern of operation and regulation for telephone and telegraph had evolved to a hierarchical model based on transport practices. Early communications networks shared many characteristics with the transportation networks they were modeled on. They were:

- generally quite slow;
- highly constrained in capacity;
- reflective of high capital costs for construction, engineering, and maintenance;
- still highly sensitive to distance and geography;
- inflexible and unchanging because of large sunk costs;
- often unreliable;
- generally inefficient;
- still oriented toward parallel, noninteroperable, specialized carriers for special purposes (such as postal, telegraph, telephone, and radio networks).

Rapid changes in telecommunications technology began just before the start of World War I, but they were not adopted by the civilian economy until the first years of peace (Solomon 1978). Experimentation with radio led to the first wireless telegraph systems. The conventional wireline carriers were frightened of traffic diversion and possible obsolescence. As early as 1907, radio telephony was demonstrated, and during World War I a secret radio-telephone link was built by AT&T between a site in Arlington, Virginia, and the Eiffel Tower in Paris, for Allied communications. AT&T's early investment in radio research and development was both defensive and quite in keeping with its conventional line of business. It had bought and further developed DeForest's Audion amplifier vacuum tube patents to prevent competitors from gaining an advantage in telephony, as well as to develop expertise in the new threatening technologies. Audion amplifiers made it possible for the first transcontinental voice telephone line to be completed in 1914; and the amplifier permitted AT&T to employ a more sophisticated wireless voice system without infringing on a number of wireless telegraph patents controlled by other firms (Aitken 1985; Archer 1938; Lewis 1991; Schwoch 1990).

But the model for radio development was much like that for wireline communications: point-to-point, hierarchical, common-carrier message systems. The technology had changed, but no one seemed to notice its key parameters were not coincident with the operative assumptions for network regulation; that is, radio was different from telegraphy or telephony:

- Its costs were relatively distance insensitive over wide areas.
- It sent communications in all directions at once.
- It was easily intercepted.

Radio's public resource was not rights-of-way or cable landings, but the ephemeral "ether"—a resource that seemed more akin to scarce riparian (water) rights than roadway or railway easements.

In the early days, however, wireless was as immensely capital intensive as landline and railroad infrastructure. Wireless required large areas of land for antenna "farms" and very expensive steam-driven alternators for producing radio waves. In fact, the total number of radio stations predicted by the prime manufacturer right after World War I was only nineteen worldwide—all of which would be transmitting wireless telegraphy, not voice or broadcasting as we know it (Aitken 1985; Solomon 1989b).

In building models for business and government policy based on the technology of the time, this lack of foresight on the potential future application of technologies was not unusual. As digital technologies pervade daily life in the 1990s, the impact and further growth potential of networked markets is grossly underestimated. For new technology to break the shackles of old-market and regulatory models is surprisingly difficult.

The common-carrier model did not fit radio broadcasting, so a new model, discussed in more detail in chapters 3 and 5, was made to fit; and the schism between one-way emanations and two-way common carriage for the public-switched telecom network will remain until there is another major technological shift. Our mental models are tenacious, until some externality finally comes along to break the mold.

In particular, the cost slope for telecommunications transmission has become steeper (that is, it is becoming cheaper faster), and processing is faster and more powerful. Transmission speeds are faster than computer processing for the first time, signifying another radical shift in the way telecom will be used.

Although both transport and telecommunications increase value by giving access (to land or to information), the easy bypass of associated real estate for telecommunications provision makes it fundamentally different. Geography plays a decreasing role in telecommunications logistics, yet it is inherent in transport implementation.

With computers on the network, the physical circuits—wires, soldered joints, switches, fibers—are much less important in determining utility, costs, and prices than the virtual or logical path that communications traffic follows. So, even though transport and telecommunications network topologies may have superficial resemblances—perhaps useful in the past for regulation and public policy—it is now imperative that the differences between *physical* networks, as typified by transport, and *logical* networks, such as the Internet and other telecommunications networks, be understood by policymakers. If we continue to follow old, unworkable analogies, we may never get away from the outdated models of regulations and laws that could cripple the nascent National Information Infrastructure.

With symbols and electrons, we can design an object in Singapore and manufacture it in California, or vice versa, without moving any physical substance between these two places. Manufactured goods no longer have to be stockpiled in warehouses or shipped with long lead times to prevent factories from shutting down because of lack of parts. Telecommunications networks and carefully designed computer programs can orchestrate the design, manufacture, and distribution of parts and deliver them to an assembly or sales point "just in time," thereby eliminating waste or ill-advised production and cutting overhead costs; telecommunications substitute for space. And with ever more sophisticated devices, teleconferencing and videoconferencing are replacing human travel. Indeed, without telecommunications, travel itself would be a nightmare of overbooked or underutilized transport, complicated hotel reservations, and luggage dispatched to the ends of the earth, as it still is in underdeveloped regions of the world.

What used to require going to a physical place—the ancient Greek agora, the trading house, the quay, the theater, and most recently the shopping center—can now be done better and more efficiently via telecom lines. Faxes and electronic mail are replacing postal delivery just as the telephone replaced the telegraph messenger boy decades ago. Not that all transport will disappear. On the contrary, transport becomes much more efficient with telecommunications, and indeed it was the early railways, and later the airways, that pushed the envelope of electrical communications because their own infrastructure could not expand without it.

In sum, telecommunications concepts of transmission and speed have weak analogies to transport; transport concepts of costs, distance, and speeds are linked in totally different ways than are those same aspects of digital, electronic telecommunications systems. There is no analogy in transport networks to the value-added offered by computer processing—the key technology that has changed telecommunications and brought about the convergence of media. If anything, for goods movement, transport decreases value because of time delays in shipment, while telecom networks increase value by their very nature.

The electronic agora—what some call cyberspace—is destined to be different in kind and extent than the physical agora and its inherent location-specific infrastructure linkages (Solomon 1991). Electronic communications need to be treated differently, and the obsolete models of transport must be dropped in favor of new paradigms that reflect the changing world.

Technology Models

It is common to find public policy toward a new technology patterned after some older form that it appears to resemble, at least at first (Pool 1983). Thus we find that common-carriage regulation of telephones (and telegraphs) was patterned on a nineteenth-century rail transport model. Nationwide regulation for the telecommunications industry did not really become effective until the beginning of the twentieth century (with the 1913 Kingsbury Commitment). The 1934 Communications Act used wording from the 1887 Interstate Commerce Act, replacing "railroad" with "telegraph and telephone" (Horwitz 1989). Radio, the new hierarchy-flattening technology, was treated like a different animal. This act, with its antiquated image of telecommunications as no different from railroads, remained the dominant law defining U.S. policy toward communications systems until 1996.

By the outbreak of World War I, following ruinous tariff wars and other chaotic corporate behavior, railroad corporations (and their banking underwriters) acquiesced to a common-carrier regulatory regime to balance its needs for óligopolistic capital formation with those of the public and the government for nondiscriminatory service and tariffs (Klein 1987; Kolko 1963). Common-carrier law also limited carrier obligations to customers due to service interruptions, acts of God, and similar transport discontinuities. And the concept of franchising, seemingly inherent to common carriage, raised the barriers to entry for competitors (Hilton and Due 1960).

The telephone carriers, after starting off in a more laissez-faire manner than even the railroads, rapidly adopted a similar common-carrier model to avoid some of the confrontations with both government and subscribers that characterized the growth of the steam railroads. The common-carrier model guided almost a century of investment, research and development, and implementation. Characteristics subsumed in that model included a "natural monopoly" for "efficient utilization of resources," with potential monopoly abuses balanced with universal service funded from cross-subsidies within this natural monopoly (Horwitz 1989; Pool 1983; Schivelbusch 1986). Eventually overriding the model was an increasingly complex system of government regulation, and some degree of self-regulation, to define the bounds of the monopoly and the services. There are of course some positive aspects of the common-carrier model; for example, it requires that service be provided on a nondiscriminatory basis to anyone who wants it. The challenge is how to achieve universal access in a world of heterogeneous networks. The digital revolution and new network architectures discussed below are key elements for achieving the goal of universal access without common-carriage regulation.

The Digital Revolution and New Network Architectures

To understand why we believe a new network architecture, conforming to the principles of an Open Communications Infrastructure, is possible and desirable, a review of the principles of electronic communications and computing systems is helpful. The key characteristics of computers and digital telecommunications networks are summarized here.

Today's computer-based, all-digital networks are as far from a railroad-like architecture as smoke signals are from satellite communications. Different, as well, are the assumptions necessary to develop a new model for public policy. The technological drivers are no longer the cost of installing copper wires, maintaining electromagnetic switches, and making physical connections—all of which had reasonable analogs in railway transport. Today the drivers are computers, photons, wireless connections, and microchip logic devices. Networks have logical paths, rather than physical paths; maintenance can largely be automated in software and done from remote locations; and amortization is paced by the rapid change in chip fabrication, on the order of eighteen months for new processors, and less for some other components. Distributed control in telephone networks—particularly Internet Protocol (IP) networks—complexity, and interoperability are other key characteristics of digital networks (Solomon 1987d).

Long-haul circuits have decreasing costs because of sophisticated methods of multiplexing bitstreams on strands of fiber and decreasing unit costs of the interface devices connected to the fiber. As the mesh of fiber has grown in density across the continent, economies of scale and new nonhierarchical routing software have given new meaning to distance insensitivity: it makes no difference how the signal gets there as long as continuity can be guaranteed and latency (the delay in signal arrival) and switching times are acceptable to the user. So true costs for transmission are the same no matter where on the net you originate or where you terminate.

But the catch is getting on the net: local connection to the long-haul network is still constrained by physical parameters and gateway circumscriptions. Indeed, the cost gradient does not rise steeply for telecom today until the last few hundred feet of local access and distribution. One obvious way around that is to use digital cellular radio technology for the last mile. We discuss that opening below.

Other key characteristics of the new digital networks that break the old molds include: variable bandwidth demands; increasing economies of scope defining interconnectivity; and channel diversity, symmetry, flexibility, and extensibility. The new model for telecom networks is *internetworking*, as exemplified by the Internet, a network of networks defined by its interface protocols instead of its corporate structure (it has none), tariffs (determined by access to backbones), physical routes (it uses whatever is available, public or private), or boundaries (whatever connects becomes immediately part of the Internet).

The final characteristic of the new networks, something very foreign to transport, and inimical to old-line telegraphy and telephony, is intelligence—on the network, but more important, at the customer end. Intelligence is defined in terms of computer processor power and the software needed to interface with the network and provide application functionality. To interwork, customer equipment must follow carefully defined protocols and standards. Interestingly, one of the reasons for the widespread and rapid success of the Internet is that the interface protocols have been built into all Unix computer operating systems since the 1980s at no extra cost to the users. Internetworking was inevitable under those circumstances (Solomon 1989d).

Because standards are so important to understanding the nature of the new networks, we address their critical role in telecommunications below.

Standards

Nineteenth-century infrastructure paradigms called for standards that are dimensional in scope—track gauge, screw threads, voltage, and current levels, as well as synchronization signals. A minor change, and system elements will not connect properly, threads will break, appliances will burn out, or get out of synch. In nature, minor changes may be regressive and the organism fails to survive or evolve further, but species that do survive appear to overcome entropy—they contain negative entropy, which is "information," as defined in communications theory. Nature handles this by dividing organisms into species, subspecies, and individual organisms; linkages are by function, animals have ecological niches, the system survives and mutates, however badly, but life does not stop completely—so far.

The interfaces between species and organisms and the biosphere can be thought of as separate layers of systems. My system interacts with your system through interfaces that our bodies understand (chemical, auditory, visual)—recognized and complex signals that communicate information of one sort or another. Sometimes the signals get mixed up and communication does not work. Usually some self-correcting mechanism comes into play to handle fuzzy signals; sometimes it does not and there are catastrophic results (Bertalanffy 1968; Solomon 1990a).

Analog versus Digital

Modern telecommunications systems are evolving to emulate what nature already knows how to do, if imperfectly at times. The principles of electrical communication are not all that different from nature: channels between objects always have some noise or extraneous energy that is both useless and sometimes causes miscommunication.

All channels are noisy, and all communications systems must find ways to overcome that noise in order to distinguish the real information, or signal, from the background noise; this is termed the "signal-to-noise ratio" (SNR). Sound is composed of molecules of air vibrating back and forth; the human voice emanates from a soundbox in the larynx that vibrates and "modulates" the air leaving your lungs, which is further modulated by your lips and tongue. On a telephone system, these molecules of moving air (sound waves) cause a device to vibrate in the telephone's microphone, and this device modulates an electric current, much like your larynx modulates the air from your lungs. The electric current vibrates "up and down," or changes its voltage in unison with the sound waves, creating a wave of electrons that flow through the wire as an "analog" of the sound waves from your mouth (Fano 1961; Hamsher 1967).

With the old, *analog* telephone systems, the distortion caused by noise and other constraints of the system was overcome by careful systems engineering by the end-to-end monopoly provider AT&T, and the human brain on the receiving end. For reasons not yet understood, people can understand other people speaking the same language, speaking in recognized accents, and speaking in familiar contexts even when there is static, missing syllables, mumbling, and severe distortion of pitch. Human intelligence is able to overcome the noisy channel (Cherry 1966; Pierce 1990).

A number of techniques were invented through the years to overcome noisy channels and amplify weak signals on analog voice telecommunications systems, but none was perfect. Eventually, after a series of amplifications of any analog transmission, a signal will degrade to a point where it cannot be used. Furthermore, most of the systems invented were quite expensive to build and maintain (Pierce 1990).[1]

Then an extraordinary thing happened in conjunction with the invention of the digital computer during World War II: a perfect "amplifier" and repeater device was developed. There was a catch, however. This device, which itself is really a subset of a digital computer, required the sound wave to be converted into binary numbers so that this digital device could process the signal. This is Pulse Code Modulation (PCM), a fundamental breakthrough in transmission technology. Not only can a PCM logic device repeat a weak signal (for there is still noise in the line, and the signal, although digital, still degrades just like analog waveforms degrade), but digital systems can be designed to repeat the signal *perfectly*, and ad infinitum.[2] Further, these numbers representing waves can be processed in computers, so the numbers can be routed or switched from one circuit to another and can be mixed in with other numbers representing different circuits. Information about the circuit can be extracted for various purposes, from billing to routing, to duplication (for a form of "broadcasting"), and even for smooth, traceless wiretapping. The catch was that computers and logic devices, and the software that controls them, were very expensive in the early days (from the end of World War II to the mid-1960s). A further catch was that the entire concept was highly classified for many years, given that it was an outgrowth of early cryptographic endeavors (Lebow 1995; Randell 1982; Solomon 1990b).

By the late 1950s, it became apparent to AT&T that the technology behind Pulse Code Modulation would be both valuable and critical in solving a very pressing problem: how to increase capacity of local loops in congested areas, such as Manhattan, without digging up the streets and installing more cable. PCM was ideal for this, because the capacity of copper wire pairs for pulses was far greater than the capacity for analog waves; and because, among many other factors, the emerging logic devices of the 1950s made it cost effective to multiplex (and demultiplex) many pulse-encoded voice channels on one wire pair without any significant signal loss. This was not true of analog devices that mixed various wave sources. Not only can Pulse Code Modulation carry voice, but data are transparent to such circuits because it is already digital; and similar techniques for converting analog images and pictures permit perfect transmission of fax and video on PCM circuits.

Digital circuits in the local loop are not yet common, but feeders to neighborhoods, the "curb," and increasingly, cable TV distribution plants are all digital. Hence, over the last decade, voice, video, fax, and data channels have become "cleaner"—with less noise, almost no loss of signal, and greater inherent capacity—driving costs of transmission down by orders of magnitude while vastly improving service. We can hear it daily when we call across the country or across the world.

Better conventional telephone circuits have had a major impact on the convergence of media, because fundamentally all forms of electronic communications can now share the same circuits in a cost-effective manner. Today we can send faxes and data on ordinary phone lines at speeds that make it cheaper than mail coast to coast, and often overseas. It is doubtful that analog technology could have become as cost effective as to make such information appliances commonplace—a hidden aspect of the digital revolution not normally appreciated by the lay public.

The other leg of the digital communications revolution has been the computer itself. The first digital computers that followed a sequential program—the stored-program machine—were developed for code breaking by the British during World War II. Coincidentally, they were designed by the same team that built the first pulse code modulation digital communications devices (Hinsley 1995; Randell 1982; Solomon 1990b; Weizenbaum 1976).[3]

A digital computer is a strange machine, different from all other machines ever invented. It sequences numbers, and by using tricks of arithmetic, some deceptively quite simple, it can emulate the process of any other machine, including any other digital computer. Clearly this is useful for counting, tabulating, and mathematics. And with sufficient speed and memory space, a digital computer becomes a useful telecommunications switch. A logical device—the computer—can transform itself to perform the functions needed for whatever task is before it. The physical device is less significant than the logical structure by which it operates (Weizenbaum 1976).

But these are old ideas. Why have we had to wait until now to get a convergence of media based on digital technology?

The Convergence of Media

World War II ended half a century ago. So why have major policy decisions over the past fifty years ignored the computer as the key integration device among media until relatively recently?

First, we needed to understand the nature of digital information, what the limits were for digitizing analog information and how it can be best transmitted efficiently on noisy channels. The answer was provided by Claude Shannon and Warren Weaver after the war, based on work done at Bell Labs on cryptographic systems (Shannon 1953; Weaver 1949).

Next, basic techniques were required to program the stored program digital computer; languages had to be written to increase efficient coding, mathematical theory had to be developed and tested, and large numbers of people had to be trained. An immense effort in computation was undertaken after the war, spurred initially by the demands of the cold war for nuclear weapons and more code breaking, and an immediate need to bolster air defense. From 1950 to 1960, the supersecret National Security Agency (NSA) funded R&D on computers to the tune of $1 billion per year (in 1950 dollars), mostly unclassified, and primarily using the National Science Foundation and the Atomic Energy Commission as fronts for grants to universities and private corporations. Converted to 1995 dollars, this figure dwarfs the amount spent today to maintain U.S. leadership in technologies that will continue to have critical importance for national security as well as for economic growth (Snyder 1980).

Many more billions of dollars were spent on air defense in that period, with a large proportion going to computers. International Business Machines (IBM) was the prime contractor for the air defense system, training several thousand systems engineers to build and maintain the network. This aided IBM's entry into commercial computing later in the decade. AT&T's Western Electric built a new wideband telecommunications network for the air defense computers, which had a major effect on linking the North American continent, from the Arctic Circle to the Rio Grande and from coast to coast with broadband microwave circuits and coaxial cable. At the end of the decade, direct-distance dialing was feasible, helped along by this cold war investment (Solomon 1990b). Still, this was not enough for a telecommunications revolution, but it certainly gave the United States a jump start. By 1958, with electromechanical telephone switches controlled by electronic computers, it was clear that from the 1940s Flowers's and Turing's vision of an all-electronic digital network might still be realized. All connections would be by bits–switched on a computer itself, not a computer-controlled mechanical switch–and the network would be transparent to the traffic. Even intelligence in the network was inferred. The message was ignored by policymakers, and it even spooked the computer industry somewhat. Just two years earlier, in the 1956 Western Electric consent decree, AT&T had agreed to get out of all other businesses including computers (for example, they made all the sound equipment for sound movies), and remain only in the core telephone business. To avoid confusion and the wrath of the Justice Department, they called computerized switching "stored-program control" (SPC), as if to suggest that no computers were involved. And on the other side, the computer industry was scared to death of government regulation, resisting talk of a "computer utility" connected online to the world as that became a reality with the introduction of time-sharing, networked computing systems (Horwitz 1989; Solomon 1991; Flowers 1976).

With such inhibitions, bad policy and bad technology are made. Any merger between industries was held back, to the point that when the FCC began its investigation of computers on the network with the Computer Inquiry I proceedings in the 1960s, both the regulators and the industry were blinded to the opportunities of a merger. Instead they bided their time with frustratingly irresolvable debates about what is "basic" service and what is "enhanced." Virtually nothing useful came out of this or the subsequent FCC dockets, Computer Inquiries II and III, because the technology was moving so fast that the definitions had become blurred even before the panels began their discussions of the subject. The idea that distinguishing between computing and communications systems was impossible, while it may have contributed to the nightmares of sleepless regulators, it did not enter into the policy discussion. The implications of the FCC Computer Inquiries are addressed in chapter 3.

Microelectronics

The development of microelectronics was spurred by the space race with the Soviet Union; chips were cheap, fast, and small enough to bury in all sorts of equipment; on the pole, in the telephone, and in switching. Intelligence could finally be disseminated without having to achieve economies of scale in a centralized location. Billions of additional dollars were spent by the National Aeronautics and Space Administration (NASA), the Department of Defense (DoD), and the Department of Energy (DoE), among other federal agencies, on microelectronics research, development, and procurement in the 1970s. The first four-bit microprocessors were introduced in 1970, signaling a permanent revolution in the growth of computing and communications power that was not recognized by regulators until a decade later (Hanson 1982).

With microelectronics, the machine counts very fast indeed. By 1994, chips from Digital Equipment Corporation (DEC) could do 400 million instructions per second; a billion per second should be achievable by the turn of the century or earlier. At these speeds, numerical programs can be written to emulate a lot of things that humans used to do, and many things that would take them eons to accomplish, assuming they could do them at all. This is not to say that the computers think, but sometimes it appears that they do.

When applied in digital telecommunications systems—in conjunction with technologies and software that digitize voice and images—microcomputers change the way the networks offer services, work internally, and even manage their own administration and maintenance. Computers permit the network itself to behave more intelligently. Indeed, intelligence is now found not only at the central office for switching, but also on the circuit, in the application, and especially in the customer's appliances and terminal devices. Allocating functionality to these various slices of intelligence is very complex, and distribution of cost even more complex. For now the customer may appear to the network as a carrier of information, and the network may appear to the customer as another customer. The convergence of computing and communications systems into an information highway that can reach across to any computer or person on the network of networks relies on the amazing power of bits. Bits are the essential paving stones of the information highway, as well as the special construction material of the information carriers riding on them (Negroponte 1995).

A Bit Is a Bit

To the network, all bits are more or less the same; all the network needs to know is its obligation to deliver these bits at various times (latency) and in various combinations (traffic capacity), to various destinations (routing), at various levels of accuracy or security. The bits that represent a television program of a sporting event must be delivered immediately, but bits of a movie can be delayed, as long as the bits line up correctly in the user device to present a complete moving picture. Bits that represent money must be delivered with absolutely no error (or at least all errors must be noticed and corrected); but bits that represent voice can miss once in awhile (causing a click on a typical PCM line) because the human ear and brain can figure out what was said anyway. Bits that are encrypted must be perfect, or the encryption system may not even begin to decode them, but to the network, they are just bits.

Charging for some bits differently than for others can lead to truly bizarre outcomes: for example, if you need tens of millions of bits per second for a video, the market probably would not tolerate any charge much more than a few dollars an hour for such a capacity or bandwidth. But if you charged on a bit basis for telephone calls the same as for a video, telephone calls would be virtually free. A bit is a bit. If you tried to charge the current going rate for telephone calls, the videos would cost hundreds of dollars per hour (Anania and Solomon 1988). As interconnected high-speed networks such as the Internet proliferate, the threat of undermining the existing price structure of the telecommunications industry becomes extreme (Bailey and McKnight 1995).

The model for telecommunications, when all bits are equal, is not the hierarchical model of telegraphs, telephones, and radio broadcasting. Something has to give. Differentiating by time of day, amount of traffic, or delay time may permit some differential pricing, but not enough to allow carriers to continue to charge by value of service or separate networking by type of service or content. If all bits are equal, and networking becomes a commodity, how then will a carrier earn a profit, much less attract capital for investment and expansion?

Value added is the only way to make a living in the future, but this now puts all carriers into direct conflict with their own customers, who are using the networks to add value themselves. This quandary cannot be solved by regulation, because regulating a network as complex as we have described would require micromanagement of the finest details. There are too many ways a computer can divert, subvert, and convert traffic to neutralize any fiat from on high. The paradox introduced by bits has to be solved by the market, with the government setting the rules to prevent barriers to entry created by the intricacies of the network itself. Access is the key word for future networks, not regulation. And access implies an understanding of how digital networks connect to the outside "real" world.

Protocols and Railroad Conductors
For all this digital technology to work, the various elements of the system must follow precise protocols or instructions for what to do with the bits and standards for interfacing the electrical and light pulses coming out of and entering into diverse pieces of customer equipment and sundry network appliances. Standards are critical for open systems, and the practices we use to set standards are neither trivial nor easy to understand. Moreover, the arcane procedures in standards setting and protocol implementation are full of traps that can just as easily close a system, giving vendors maximum advantage over users to the detriment of a universally accessible National Information Infrastructure. They can also lead to massive distortions in investment.

Like the telephone, the early telegraphs using Morse's "dit-dah and space" code also depended on humans to figure out which letter was being represented; the telegraph operator had to determine where the dits started and the dahs ended, and then write down the letter it must be. This was a slow process. By the end of the nineteenth century, a number of printing telegraphs were in use that automatically translated the telegraph code and displayed the text on paper. Most were based on a code invented by Emile Baudot, whereby each character was preceded by a signal that told the telegraph it was beginning, and another signal that indicated the end—somewhat like the RETURN key on a computer keyboard that tells the computer to process what it just received. The Baudot system, still in use on what remains of the world's Telex system, was grossly inefficient for higher-speed communications needed by even the earliest computers. So when computers started to be interconnected in the 1960s, for airline reservation systems, air defense, and access to time-sharing systems, better methods were necessary to cut costs and provide 100 percent accuracy, because if one bit was missing the machines would fail.

Furthermore, to lower circuit costs, early computer interface standards needed to multiplex or mix bitstreams from different sources, including control channels. This has become the rule for virtually all telecommunications circuits and networks today. A collateral principle was designing the signaling protocol to handle a variety of noisy channels. Unlike an analog voice channel, where the human ear/brain combination does most of the work of interpretation at either end, a digital system depends on special mechanisms (some of which are in the software programs being served by the bitstream) to tell it where the detected pulses belong in the larger scheme of things. Just transmitting bits in the form of light or electrical pulses on a circuit accomplishes nothing; protocols are necessary to receive these pulses, and these protocols can be very complex given that they have numerous functions to perform besides mere detection.

The model of multiplexed channels and the concept of synchronous bitstreams, moving in chunks or "packets," came together in the late 1960s in a radical conceptual shift: the packet network—another major policy change that was ignored for a very long time. In 1968 attempts to build packet networks began in earnest. In the United States, the Defense Department's Advanced Research Projects Agency began design of the ARPAnet, and in France IRIA began the Cyclades network in 1973, both seminal in developing the standards and protocols used today in digital networks of all types, from local area networks (LANs) to so-called Integrated Services Digital Networks (ISDN) for the telephone system. Parallel efforts were launched in 1969 by the worldwide airline reservations system and TYMNET, a commercial timesharing service, to develop packetlike networks and protocols (Lebow 1995; Mathison 1970; Solomon 1991).

The Convergence of Media

Since the 1970s we have been hearing that the boundaries separating telecommunications, computing, television and radio broadcasting, publishing, games, telephones, faxes, and e-mail are becoming blurred. What this means is simply that all human information—what we see, what we hear, and someday even what we taste, smell, and feel—can be converted into a form that digital computers can process, store, and display. Moreover, this information can be readily transmitted almost instantaneously anywhere on the planet (and beyond) at very low cost, or at least, very cost effectively.

The operative word is "process." The stored-program computer, with clever software, can dig through mountains of bits very quickly: find contextual connections across data stored in different locations; make inferences not noticeable to people; find information where humans see only fuzz or hear only a buzz; restore data thought lost permanently; fraudulently alter records so they cannot be detected; and totally change our view of how we communicate.

The packet networks, coming after three decades of development in digital telecommunications systems, enabled the development of new network architectures, tariffing principles (based on volume, data rate, or latency of information flow rather than distance), and radically different network uses. From the ARPAnet research, a number of fundamental standards concepts evolved, which are expressed generally in its Transport Control Protocol (TCP) and later Internet Protocol (IP) that define not just transmission standards, but an entire system architecture (Comer 1991). The current standards debates in telecommunications derive more from these efforts than from digital telephony or telegraphy itself; but the older forms continue to constrain the newer ones (Solomon 1990b).

Layers

As noted above, the desire grew to build network protocols that were rational, flexible, and well understood to deal with noisy channels. These technical reasons, as well as the economic need to multiplex different channels to reduce carrier costs, led to transmission-efficient techniques for synchronous transmission of "frames" of bits. This eventually ushered in a layered approach to a total network system, where each lower layer nests into the layer directly above, so that a layer is only cognizant of the layer below and the layer above. Everything else was "baggage" to be delivered. An analogy could be made to a conductor assigned to one railroad car on a long train of cars. Conductors know who is in their car, and even know where each passenger is to get off. This is the level "below" the conductor. Conductors can walk from one end of the car to the other, but not to the next car. They "know" their car is connected to a train (which is the level "above" the car), and they also know when the train stops and who gets off where, but they neither know nor really care about the rest of the train.

The information about the passengers is self-contained within the car, although the car is part of a train that is taking the passengers to their destination. Replacing the train metaphor with a data packet, we have an inverse pyramid where the lowest layer connects to the actual circuit, and only the data in the center of that layer is important to the circuit, and to the data in the layer immediately above it. All the rest of the data in the layer is baggage; the packet train only knows to deliver the baggage-data somewhere, but cares little about how the data is packed or arranged. Indeed if the data does not get there intact, it is not the problem of this layer. Some other layer has the responsibility to reorder it, discard it, or change it. If every part of the railroad had to worry about every other part all of the time, the resultant overhead would cause so much congestion that nothing would move. This is the basic philosophy behind designing a network model that determines reference points in a total system architecture for different network functions.

These concepts will be essential for the development of open systems that are underpinned by functional standards. Furthermore, these open systems concepts, and the development of standards and protocols to implement them, enable the development of an Open Communications Infrastructure of information highways that allows the dissemination of different types of information owned and distributed by a variety of institutions and individuals.

In computer-communications terminology, the concept of a "protocol" is a constituent part of an interface "standard." It is "a format and a set of procedures that are *commonly agreed to* for the purpose of achieving communications. Layering of protocols is the result of a *stratification of function* among parts of the system" (Stallings 1987).

Network Theory

A network implies connectivity among links, nodes or junctions, and users or user equipment. Except for specialized, narrow-purpose networks, from the user's point of view, the easier the connectivity, and the more of it, the more the network is worth. More connective utility can offer economies of scope to the user; and when economies of scale permit the spread of declining unit costs over a larger system, the network provider gains greater market share. Unfortunately, short-term incentives tend to obscure this fact for the monopoly provider, who would rather keep the savings as profit—unless competitive forces, or as in the past, regulation causes the provider to lower prices, lower barriers to access, or offer more and differentiated services.

Networks come in varied architectures. The current telephone model of a hierarchical structure, with dominant nodes or junctions connected by primary trunk lines fed, via subsidiary nodes, by secondary and tertiary circuits or branches—that is, a tree pattern, or more accurately a double star—evolved physically from the railway engineering practice and organizationally from military, line-and-staff command and control (Pool 1977).

The process is no longer centralized; in fact, it can never be again, because the processors are too cheap and are getting cheaper, and work best when networked. With distributed control, the genie is out of the bottle and the bottle of centralization is broken, and there is no technological or political imperative to fix it. What we now have are many interconnected, general-purpose networks and general-purpose machines that can emulate any process. The Internet is just that, a network of networks. The protocols we discussed for internetworking have been built into the operating systems of virtually all minicomputers and work stations for more than a decade; it was a trivial decision to make because it was necessary to interconnect machines on a local level anyway. The next generation of personal computers and laptops will all have internetting software—again a trivial decision given the user's need for peripherals.

All the Internet does is find a connection across public, private, or specialized networks. The connection is paid for by the user or the user's sponsor, and it is not expensive. This network of networks has grown to an estimated 50 to 80 million users worldwide. The so-called public subsidy for the U.S. backbone connection amounted to about $0.50 per user per year before commercialization in 1995, much less than the general perception of heavy government subsidy.

The user pays about 90 percent of the total costs of internetting—for a workstation or personal computer (PC), software, the local nets, and external connections—most of which would be paid by the user in any case for office, campus, or general communications. The externalities of the Internet have profound implications for what a network will be in the twenty-first century.

With the introduction of the Internet, the nature of networks has entered a new, permanently altered state of decentralization. Statistical sharing, positive network externalities, and interoperability are all benefits realized by distributed computing through the network of networks. Internet software and internetting protocols today offer sound—for voicemail, voice conferences, or audio broadcasting—and video. Sophisticated processing at the user end can grab text, moving images, and data from multiple sources with ease; information dissemination is no longer the exclusive province of gatekeepers, newspapers, broadcasters, or government functionaries. This is what boundary blurring really means: access for all.

But there is a catch. No matter how sophisticated the software for display is, you still have to be literate; no matter how good the graphics are for charts, you still have to understand statistics and mathematics; no matter how slick the video might be, you still have to know how to put facts in context. This is not a case of "computer literacy," but rather of education. And you have to get access to the network of networks, and that still costs money.

Most important, to assure that the networks interconnect and reach as broad a base and population as possible, the boxes and appliances must follow precise, if not rigid, protocols. Because no one has a clue as to which applications are going to be needed or become successful, a set of general-purpose networks built with Open Communications Infrastructure principles will provide the only way to reach the broadest possible customer base. OCI will give us a series of interoperable platforms on which we can build as yet unknown applications with less risk than the networks it will replace: common-carrier constrained, public-switched telephony and telegraphy grids, or broadcast-oriented entertainment architectures.

From Convergence to Divergence

With bits being bits, the marginal revenue for carriage will be small; only adding value will make sense for carriers. But the transition will not be simple.

As we have noted, the public-switched telephone network has evolved— with critical help from computation and processor microelectronics— from a set of hierarchical structures, interconnected by human operators and slow-moving electrical switchpoints, to one that today carries voice, video, and data rapidly, accurately, and relatively cheaply (compared to yesterday when it was expensive or simply technically impossible). But the public-switched telephone network is still optimized for two-way voice, not data or any of the variations that digital telecommunications engenders.

From Physical to Logical: Digital Network Architecture for an Open Communications Infrastructure

The physical connections of highways or electromechanical telephone switches are, in modern times, anachronistic. They are a misleading and inappropriate metaphor for communications infrastructure. Computers are inherently communicating devices, a fact that has been overlooked at times by computer manufacturers, to their own peril. We have shown how the logic of computers has already spread throughout the telecommunications infrastructure, and continues to spread across the technical infrastructures of many converging industries. We have shown how one can find analogies between the technology and organization of railroads and the network of networks of our time: the Internet. Although the structure of railroad networks does offer some useful metaphors, it too has been superseded by the logical nature of computer networks—which powers all modern telecommunications systems. It is in the workings of the Internet that we can begin to discern the technical architecture and design principles needed to enable development of a self-sustaining Open Communications Infrastructure. Whether the distributed nature of the Internet is a source of strength or a potential weakness in the face of state power remains to be seen. Our hunch is that the state will find the Internet especially difficult to control, which has both positive and negative consequences. How the state has reacted over time to evolving network infrastructures is the subject of the next chapter. The time line in table 2.1 is intended to help orient readers to the myriad developments in technology, law, and policy that have shaped the communications infrastructure.

Table 2.1
Master time line

1791	Hamilton's report on manufactures
1808	Gallatin's report on roads and canals
1824	*Gibbons v. Ogden* establishes federal authority over interstate commerce, thereby preventing state monopolies
1830	B&O Railroad—first U.S. common carrier railroad
1833	Daily newspaper
1836	Babbage's prototype digital computer (analytical engine)
1837	Photography invented
1844	Morse telegraph
1851	Illinois Central Railroad—first federal land grant railroad
1858	First working transatlantic telegraph cable
1861	Civil War starts
1861	Land Grant Telegraph Act
	First transcontinental telegraph
1861	Reis demonstrates crude telephone in Germany
1862	Morrill Act—Land Grant colleges
1863	Land Grant Railroad Act
1865	Civil War ends
1865	Maxwell describes mathematical theory of electromagnetic (radio) waves
1866	First successful Atlantic telegraph
1869	Transcontinental railroad completed
1876	Telephone invented
1877	*Munn v. Illinois* established governmental right to regulate firms operating in "the public interest"

Table 2.1 (continued)

1877	Phonograph invented
1883	Hertz demonstrates radio waves
1885	American Telephone & Telegraph formed
1885	*Corsair* cabal—J. P. Morgan settles railroad wars
1887	Interstate Commerce Act regulates railroads
1888	Bell patent cases settled
1890	Sherman Antitrust Act
1891	Edison demonstrates motion pictures
1895	Marconi wireless
1896	First commercial automobile
1901	Marconi transatlantic radiotelegraph
1906	Fessenden's radio broadcast of singing voice
1910	DeForest's triode vacuum tube
1910	Rosing in St. Petersburg, Russia describes working television
1910	Mann-Elkins Act extends Federal regulation over telephone and telegraph
1913	Kingsbury Commitment halts AT&T expansion and divests Western Union Telegraph
1914	Transcontinental telephone
1914	World War I breaks out in Europe
1916	Farnsworth invents modern electronic TV
1916	Railroads nationalized
1917	United States enters World War I (Zimmerman telegram)
1918	Secret transatlantic radiotelephony
1918	World War I ends
1919	AT&T nationalized
1919	Radio Corporation of America formed
1920	AT&T denationalized
1920	WEAF/KDKA radio broadcasting begins
1921	Railroads denationalized
1921	AT&T given limited rights to continue mergers
1922	Radio networks formed
1925	First crude mechanical television signal broadcast in the United Kingdom
1926	RCA/AT&T patent cabal
1926	NBC founded
1927	Mechanical color television/picturephone demonstrated
1927	Federal Radio Commission
1927	Transatlantic commercial radiotelephony
1928	Sound movies
1929	Great Depression begins

Table 2.1 (continued)

1930	All-electronic television demonstrated
1934	Communications Act
1935	BBC begins regular TV broadcasting
1935	FM radio
1936	A. Turing and V. Bush independently describe mathematical properties of stored-program computers
1939	First commercial TV broadcast in the United States
1939	World War II begins
1940	RADAR
1940	AT&T demonstrates remote computation with teletype and relay calculator
1943–4	First stored-program digital computer—Colossus—secret British code-breaking machines
1945	World War II ends
1946	Television reintroduced with NTSC black and white higher-resolution standard
1947	Transistor invented
1949	Antitrust case filed against AT&T
1950	Cable television as master or community antenna
1953	IBM 701—first production Mainframe Computer
1953	All-electronic color television system adopted in United States
1955	Start of computerized air defense system—networked computing in real time
1956	Modem invented
1956	First Transatlantic telephone cable
1956	Videotape demonstrated
1956	Final Judgment in AT&T antitrust case restricts Western Electric to telecommunications
1957	Sputnik—first satellite
1958	AT&T begins experiments with electronic switching
1959	Integrated circuit
1959	Above 890 microwave case
1960	Laser invented
1960	Pulse Code Modulation introduced
1961	Nationwide direct distance dialing (DDD)
1962	Comsat
1964	Picturephone I
1965	Hush-a-phone
1965	Geostationary satellite
1969	Microcomputer on a chip
1969	ARPAnet—first packet network

Table 2.1 (continued)

1969	Northern Telecom—first commercial digital switch
1969	Fiber optics demonstrated
1975	Satellite-based cable television
1976	VCR invented
1977	Altair—first personal computer
1980	First long-haul fiber-optic line—Washington, D.C., to Boston
1982	Breakup of AT&T modifies 1956 final judgment
1984	Cable Act
1988	Internet expands with NSF backbone
1992	Cable Reregulation
1994	World Wide Web expands functions of the Internet
1995	Internet backbone commercialized
1996	Telecommunications Act

3

The Network and the State

Although our call for a bold proactive federal policy for an Open Communications Infrastructure reflects a fresh approach to a difficult problem, it is certainly not unprecedented. A cursory historical review may leave one with the impression that the private railway and telegraph systems, for example, had little to do with federal policy. But that was not at all the case. The role of government was, in fact, pervasive. This chapter offers a brief glance back to refine our policy proposal by drawing lessons from the successes and failures of the past, with special attention to the delicate dynamics of public and private incentives inherent in the nature of building and maintaining public networks.

We address two historical themes here. Although we might deal with the two sequentially, we find that they are intertwined. The first is the need to acknowledge the pervasive involvement of the federal government in the development of privately owned public networks. As the ideological pendulum of American politics swings back and forth, we find ourselves in the 1990s at the peak of enthusiasm for the wisdom of private capital and at the same time at the peak of skepticism about the capacities of national planning and federal policy. Our concern, quite frankly, is that this is an awkward time for the pendulum to swing. As we move now to outline the architecture of communications infrastructure for the twenty-first century, we appear to be paralyzed by the thought that federal government involvement might distort incentives, distract private capital, and attempt to pick winners among competitors in the market. Such a view is naive. As we demonstrate, federal policy through the funding of research and development, setting of standards, allocation of spectrum, and a mix of inducements and restrictions on market entry and exit continues to be central to the business of building and maintaining public networks.

To profess that telecommunications and broadcasting are simply the domain of private enterprise is to ignore history and put the future in jeopardy. The risk is great that new monopoly positions will arise as a result of neglecting the public interest in communications and information infrastructure. As we show in this chapter, history is replete with examples of abuse of dominant positions in critical network markets and infrastructures. We are not saying that the marketplace is bad or that would-be monopolists are evil, only that the marketplace in general and dominant firms in particular will attempt to abuse the small farmers of our day—small businesses and residential customers—as surely as the railroad owners took advantage of their strategic position to extract monopoly rents from captive customers a century ago.

Our second theme concerns a historical shift in the way in which the public and private domains interact. This theme is central to our conclusions about the need for an Open Communications Infrastructure. We note that in the earliest generations of physical networks—the railways, telegraph, and early telephone systems—there was a successful program of closed and rigid standards and common-carriage regulations to guarantee equitable access and reasonable tariffs. Thus we can trace the history of the common but ill-designed railroad gauge of 4 feet, $8\frac{1}{2}$ inches through the Middle Ages back to the actual measure of Caesar's original measurement regulations for the wheels of animal-drawn carts in ancient Rome. But as the architecture of advanced digital electronic networks moves to a flexible geodesic structure, a new paradigm of control is necessary. The nature of the technology, we argue, enables competition and flexibility to replace the rigid regulation of standards and access as the best means of ensuring that the public networks are responsive to the public interest. To cling to the old paradigm and vestiges of the old rules is not to protect the creative initiatives of the private market, but rather to distort them.

To Make All Laws Which Shall Be Necessary and Proper

The U.S. government has variously planned, sponsored, and stimulated development of infrastructure from the beginnings of the Republic, particularly the transportation and communications systems. Governments in most countries have been similarly involved in creating infrastructure, some directly designing, financing, and operating systems, and some indirectly with less planning.[4] Overall, the United States has followed the latter approach. The next section shows how relations between the public and private sectors have changed over time. First, the courts had to justify giving public powers to private firms. Then, the public sector had to justify public power over private contracts for infrastructure rates and services.

The Constitution defined the powers of the federal government to control and regulate commerce between states and to facilitate commerce by providing for a communications system. To avoid repeating the failure of the Articles of Confederation dating from 1781 (Nevins 1945), the Constitution granted the federal government the right to regulate commerce among the states. In the case of communications in general, the right of the federal government to establish post roads and post offices is explicitly noted. The founding fathers saw these powers as integrative forces for the federal union. They were also quite aware that they could not anticipate the nature of the technologies that would develop over time. In essence they wrote an "open" constitution, one that could evolve as the details evolved and could be extended if and when necessary.

The right of the federal government to preempt the states in matters involving commerce has been contentious from the beginning; and indeed on the slavery issue, the controversy over "nullification" of federal law by the states, was a factor leading to the Civil War. But the Supreme Court, in one of its most important decisions establishing the meaning of the Constitution, *Gibbons v. Ogden* in 1824 (termed the "emancipation proclamation of American commerce"), provided the basis for all subsequent legislation that contributed to making the United States the integrated nation-state it is today. The Gibbons case was about a patent monopoly for the steamboat, a monopoly which claimed that the states could concentrate rights to all steamboat navigation in the one firm that held Robert Fulton's patent. Steam technology was so superior to wind and animal power that had the view of the monopoly prevailed—as the historian Burton Hendrick wrote during the New Deal when the question of infrastructure monopolies was once again high in the public consciousness—"the states of the American Union would have become individual entities indeed, constantly shutting out each other's citizens, engaging in everlasting commercial war; and the greatest privileges of civilized countries, those of transportation and of intimate, easy circulation, would have remained the monopoly of a few powerful groups" (Hendrick 1937).

This theme of monopoly control of key infrastructure is a common one in American history, one to which we return every time a new powerful technology is introduced, and one that will no doubt arise again in the near future with the development of multimedia and other new forms of electronic communications. In fact, the issue of monopoly control arose right from the introduction of the telegraph in the 1840s and the subsequent expansion of steam railways in the 1850s. These became significant economic and political forces in interstate commerce, leading to further constitutional crises over sovereignty and control of strategic technologies (Beniger 1986a; Cummings 1937; Lindley 1971).

The telegraph was immediately perceived as being different from roads and waterways because of its electrical and instantaneous qualities. However, given that there were no models to help the government deal with the telegraph, the railroad model, which was developing at roughly the same time, was initially applied. In the United States, resolving the ownership of the telegraph patents proved to be a long and almost intractable battle; for who controlled the patents could determine who would be in the telegraph business and thus who would control certain critical information (Lindley 1971). Whoever controlled the telegraph signals wielded immense power (Thompson 1947).

A decentralized government and a lack of capital made it difficult for any one federal agency to implement changes in infrastructure. Private enterprise filled the vacuum with capital provided by European investors and with other resources (timber, coal, iron ore, labor, and inventive genius) allocated domestically.

Control and Common Carriage

Granting governmental powers to private firms for the right of eminent domain—that is, the right to condemn land for a "public purpose"—was the direct forerunner of the legal concept of "in the public interest." This idea is fundamental to powers of rate and service regulation exercised over, for example, telecommunications, up to the present. The privatization of governmental functions was essential if private enterprise was to have *control* over the complex technological and economic infrastructure necessary to implement and operate steam railways, which evolved into a much more complicated system than had the privately owned canals and toll roads. The concept of control through the exercise of government power became an important underpinning for later telegraph and telephone developments in the United States; but it was not just a one-way street—in exchange for taking private property, the courts held that fair value must be paid, that all users must be treated equally, that carriage could not be refused (common carriage), and that the government retained the right to regulate charges and access to ensure fairness.[5] But the public nature of private enterprise was settled early.

Because the legislature permitted the company to remunerate itself for the expense of constructing the road, from those who should travel upon it, its private character is not established; it does not destroy the public nature of the road, or convert it from a public to a private use. (*Bloodgood v. The Mohawk and Hudson Railroad Company* [1837])

Because of their interaction with land tenure and property rights, railroads in the United States and elsewhere were guided, financed, cross-subsidized, and ultimately regulated (or nationalized) by their respective governments.

In the United States, land grants and direct appropriations built the four principal railways that had crossed the Alleghenies by 1840, as well as numerous local lines in the North and South (Federal Coordinator of Transportation 1938). The same policy led to New York State's financing and operation of the Erie Canal in the 1820s and to the federal government's expensive and unsuccessful effort to build a national road system prior to the maturation of steam railway technology after the War of 1812. Indeed, the first national road plan developed by Secretary of the Treasury Albert Gallatin, in 1808, was seen as a fulfillment of the Treasury's role as implied in the Constitution to provide revenues for the federal government. Tolls were to be collected, and the road system was intended to open up the Western Reserve (Ohio, Indiana, and Illinois) to colonize the economic base of the new nation (Goodrich 1960; Locklin 1954; Ward 1986).

Public-sector attempts at road and canal building collapsed in the panic of 1837; and a century later, this period of financial disaster was still being used as a reason for government to keep out of infrastructure development (Coman 1930). Attempts to reinstate the road and canal programs after the financial collapse were undermined by the rapid impact of steam technology on the railways in the United States, Great Britain, and Europe (Briggs 1982).

In the nineteenth century, railway and telegraph monopolies were based initially on patent rights, then on franchises. Eventually, they relied on exclusive contracts between shippers and shipping consolidators. Telegraph monopolies also relied on exclusive contracts between the owners of the wires and the owners of the rights-of-way (Lindley 1971; Rhoads 1924). Contracts were harder to overturn in the courts than monopoly grants (Grodinsky 1957; Horwitz 1989).

Governmental Aid to Transport

Railways could not have developed as they did without government intervention to reduce the risks entrepreneurs faced when preparing to invest enormous sums of capital in infrastructure. The form of intervention, land grants, provided rights-of-way, as well as the land to be sold. Investing in railroads could be viewed as speculating in real estate. With the recognition of the importance of a transportation infrastructure for economic development, however, the risks of this speculation were reduced. Almost by definition, land near a railroad was worth more than land far from access to transportation. Whether new forms of inducements are needed for investment in communications infrastructure for the information highway is a subject of some debate in the 1990s. Here we will simply establish the historical precedent for this debate, in terms of the development of government aid to transport systems over the past 200 years.

As early as 1802, Congress instituted what has become a basic practice in the development of American infrastructure by providing Ohio with a land grant for highway building (Goodrich 1960; Rae 1979). From the

1850s to 1880s, local, state, and federal governments subsidized railways mostly with land grants, but also (especially in the case of the Pacific Railway Acts) with loan and bond guarantees, military surveys, and direct cash disbursements. These techniques were used to colonize the Missouri-Mississippi basin (starting with Illinois) and subsequently the transcontinental links across the Rockies (Gates 1934). These were gargantuan projects for their day.

Modeled after the Illinois Central land grant of 1850, the 1861 Land-Grant Telegraph Act financed the construction of Western Union's transcontinental telegraph lines during the Civil War. This action was considered a military necessity to keep California in the Union (and to keep the British out of both California and the Oregon territory). The Western Union telegraph predated the transcontinental Union Pacific/Central Pacific railways along the same route by seven years. With the help of the U.S. War Department, Western Union gained near-monopoly status by the end of the war by securing control of Confederate-owned links in the North, and other exclusive military contracts. Armed with postal contracts and other government moneys, Cornelius Vanderbilt then merged Western Union into his new conglomerate, the New York Central and Hudson River Rail Road. Vanderbilt's firm was to become the major goods, passenger, and telecommunications carrier in the post–Civil War years (Thompson 1947).[6]

There is no question that the railways were essential in integrating the West and holding the country together. They formed critical military infrastructure during the Civil War and the Indian conflicts that followed, as well as during World Wars I and II. In the immediate postwar construction boom, the role of the railway companies and Western Union at the end of the Civil War had engendered antimonopoly sentiment among farmers, who saw the railways controlling their incomes by manipulating the price of their products through cartels, and among small-town merchants, who felt similarly about the way in which goods were being distributed. But the telegraph monopoly inspired federal legislation first. John Sherman, prior to his work on the Sherman Act, introduced the first anti-Western Union act in 1866 in the guise of extending the Land-Grant Telegraph Act to provide resources for competitors (Thorelli 1954). Ironically, the Vanderbilt-controlled Western

Union became the primary beneficiary of the 1866 act, using it as collateral during its negotiations with the czar for the ill-fated trans-Siberian Russian-America route (Thompson 1947).

Following two decades of acrimony, Congress passed the Interstate Commerce Act in 1887. By then the political pendulum had swung over half a century. In the 1820s, courts justified giving governmental powers to private firms to build canals and railroads; but by 1887, the government was recapturing many of these public rights. The landmark Supreme Court case, *Munn v. Illinois* in 1877, established that the government could *regulate* private business in interstate commerce under the commerce clause (Horwitz 1989; Locklin 1954; Thorelli 1954). The case is instructive because it had less to do with private property than with public order. Munn owned grain elevators in Chicago and charged exorbitant rates for his stored grain during a period of starvation caused by crop failures; riots ensued and Illinois forced the grain to be sold at fair prices. Sometimes it takes an extreme situation to change an old model.

In the 1870s, as urbanization intensified, local governments became more involved in the development of transport infrastructure. Public transport in the largest U.S. cities was influenced by public planning, development of legal rights of eminent domain, regulation and franchising of public utilities, and often direct financing of capital-intensive technology. New York's rapid transit system is a case in point. The development of the automobile shifted resources and started another technology push, although not in the manner generally supposed. Paved streets and paved highways preceded the growth of automobile ownership as a result of three forces: the burgeoning popularity of bicycling in the 1890s (leading bicyclists through what was known as the "Good Roads Movement" to pressure governments to improve roads) (Flink 1970); the growth of suburban telephony and rapid suburbanization, which followed the electric streetcar at the turn of the century; and the lobbying efforts of the railways for government to provide better access to their railroad networks (Goddard 1994; Hilton 1960).

The Good Roads Movement was quickly co-opted by the railroads at the turn of the century as a way to rationalize farm subsidies, which were important to maintaining traffic in agricultural areas. With 80 percent of the nation still living and working on the farm, there was little opposition initially, at least while the roads were being traversed by horses and wagons. In 1916, Congress enacted the first federal highway act since the early 1800s. It is noteworthy that the act called for R&D for highways, that is, an early direct federal investment in technology for transport infrastructure.

Railway opposition to highway development can be dated from the immediate post-World War I period, following the successful deployment of the motor truck by the U.S. Army in the European theater. The states embarked on massive programs to pave highways. Most of the 2 million miles of primary road that still defines the highway route system today was initially paved during this time (Goddard 1994). In the late 1930s, the federal government's Works Projects Administration (WPA) extensively rebuilt about 100,000 miles of primary system, as well as critical links over new bridges, tunnels, and dams. Federal loans and grants were used for a variety of projects: completing the Pennsylvania Railroad's electrification from New York to Washington, D.C., and Harrisburg; building much of New York City's independent subway system, Chicago's subway, the Triborough Bridge, Golden Gate Bridge, Oakland Bay Bridge, Hoover Dam, Tennessee Valley dams and power plants, and building the first modern highways, which were the precursor of the interstate system.

Superhighways received critical government assistance in the 1950s and 1960s. The interstate highway system was first proposed as a public works program by Franklin Delano Roosevelt in 1938. By the time it was completed more than fifty years later (after two decades of construction), the routes followed the general pattern laid out in the 1930s, but the capacity and detailed design of pavement and access points had changed radically. R&D, mostly paid for by the federal government, accounted for this alteration. Neither R&D nor actual route planning and land acquisition began in earnest until Congress changed financing from a fifty-fifty federal/state split to a ninety-ten split with the Interstate and Defense Highway Act of 1956. (The word "Defense" was dropped after passage of the act.) Lobbying came from the highway construction and automotive interests; and financing was provided by increasing the gasoline tax as a proxy user fee. Even though toll highways had been built since the 1930s (some even privately financed), there was no way to justify a public highway system, with an equitable penetration of roads, to the bond market.[7] This was the equivalent of free, *universal* telephone service.

Realpolitik dictated the national highway layout as it had railroad planning a century earlier. The interests of many different constituencies had to be considered. To justify construction of complex multilane and multilevel freeways in Chicago and Los Angeles, the federal government had to fund six-lane freeways in Montana as well as long tunnels in Colorado.

Infrastructure Models

The history of transport cannot tell us how telecommunications technology will develop or how to design Open Communications Infrastructure; but understanding structural relationships in transportation may help. The railroad and the telegraph developed symbiotically (Beniger 1986a; Thomson 1947). The railroad could not have covered a continent, or even a thirty-mile link, without telegraph signaling to prevent accidents. The telegraph was also needed to allocate scarce rolling stock and other resources. Conversely, telegraphy required transport for maintenance and resource development.[8] As we noted in chapter 2, although telecommunications originally followed transport architectures, the computer changed this. The model used to describe and regulate the transportation of goods and people has been generally applied to the movement of information as well. The 1934 Communications Act is a surprisingly close copy of the 1887 Interstate Commerce Act, with the word "railways" replaced by "telephone and telegraph." and references to radio communications appended as if there were no connection between wireline and wireless telecommunications.

Rails, telegraph, and telephone all developed as distinct architectures (trunk and branch or loop models), and as monopolies (essentially one carrier and one service, except at major network nodes). Although telecommunications traffic patterns, media, and human interfaces have evolved radically over the decades (enhanced by decentralized management and production), the railroad transport model is still used to describe telecommunications carriage, as well as other distributive industries, such as broadcasting and film (Solomon 1991).

The new model of network architecture more closely maps evolving digital technologies, and has profound implications for appropriate regulatory models as well. We explore below a number of technologies and related regulatory models that arose around them, beginning with rail transport, to develop a keener sense of the economic history of transport and communications network architectures. As we show, the choice between open and closed approaches to network design has long been with us, but continues to change in subtle ways as the technology and economy evolve. It should be evident by now that our orientation is toward open architectures, but we concede that in the right time and place closed, proprietary architectures are not only helpful but necessary.

Closed Models: Control of Railroad Standards

The first vehicles to be guided by rails were mining carts in the sixteenth century. They were pulled by draft animals (and sometimes men) along paths lined with L-shaped wooden slats to reduce friction. (The wheels had no flanges or raised edges to press against the rails as they do today.) The guided vehicle concept has its origins in the boats that were transported in ancient times over the Isthmus of Corinth in Greece, guided by grooves cut in a stone path. By the early 1800s, this technology evolved into "tram roads," with rails made of iron strips nailed to wood and embedded or depressed in the public road, which any wagon could use that fit on the rails. With the coming of steam traction, private steam "carriers" maintained the rails. Fees were charged for use—a variation of an older transport concept found in toll roads and canals. Quite rapidly, this process evolved into a system of iron roads especially built for the steam carriages (Encyclopedia Britannica 1911).

The question of the appropriate mechanism for control came into play early on, with the application of a standard for track gauge; the profile of the wheel flange plus the width (or gauge) of the rails became critical. Carriage clearances (the loading gauge), the weight of the vehicles, and numerous other details began to define guided transport systems, which were still small operations until the mid-nineteenth century.

Access control required administrative mechanisms to prevent collisions, handle congestion, and apportion costs. So when the first true steam railway system opened in 1830 (between Manchester and Liverpool, England), the railroad model had evolved from a partially paved, guided path in the public road to one where the railroad agency had complete control over the vehicles that ran on its private rails. Although some railroads permitted customers to provide vehicles (but almost never the locomotives), as long as they met whatever standards had been set for gauge, etc., it was not generally encouraged. (We note that this is an area in which the information superhighway metaphor fails us—for unlike the highways, but like the railroads, the controllers of information networks are likely to be able to control access in many different ways.)

It is not a coincidence that the idea gained acceptance at a point in time when railroads reached a critical mass of technological maturity in steam power and iron mongering. By mid-century the idea that the rail path provider controlled access and use, as well as set tariffs and standards, had become politically acceptable. The roads were now poised for take-off along their growth curve. It is hard to imagine that their phenomenal growth could have taken place without such control mechanisms that lowered capital risks and simplified management of innovative and very complex operations (Horwitz 1989; Mazlich 1965; Schivelbusch 1986).

Total *operational control* put the service provider in a powerful strategic position to determine further evolution. In time, operating standards were established under law in most countries, along with new political and economic rights providing de facto barriers to potential competitors wanting to enter railway markets. The antimonopoly battles of the latter half of the nineteenth century usually began by addressing abuses attributed to the railway entities. Eventually the motor truck and automobile changed the transport system, but underlining a lesson oft repeated for the telecommunications industry, the railroads were slow to begin working with competitive forces that were no longer under their technical control (Goddard 1994).

Although the details pertain to transport, the outline of the railroad standards scenario is similar to that of telecommunications in today's marketplace. The role of standards in encouraging investment in new technologies and the need for interconnection in achieving market growth provide perhaps the most significant lessons we can learn from the development of railroad standards. Given that the political resolution of railroad standards set the stage for similar interconnection infrastructure over the next 150 years, it is worth sketching the elements of the principal railway standards issue—track gauge (Taylor 1956).

As we noted earlier in this chapter, as an accident of history most road carriages in the Middle Ages inherited the old Roman cart gauge of approximately 4 feet, $8\frac{1}{2}$ inches. Julius Caesar set this width under Roman law so that vehicles could traverse Roman villages and towns without getting caught in stone ruts of differing widths. Over the centuries this became the traditional standard. It was also the predominant track gauge for the tram roads and some of the first steam railways.

Preliminary R&D in the early days of railroads indicated that vehicles would have greater stability with wider gauges (and hence could handle greater loads, thereby increasing operating efficiency). In the 1840s, Isambard Kingdom Brunel, the famous British engineer, built his Great Western Railway from London to the West Country on a 7-foot gauge; the New York and Lake Erie Railroad in the United States was built to a 6-foot gauge. To this day, although the track gauges have been converted to the Roman standard, these trunk lines have superior loading gauges and can carry traffic restricted from other routes.

Although the use of a universal standard railway gauge has obvious and significant advantages for equipment manufacturers and vehicle interchange, it has proven extremely difficult to establish over the past 150 years. A wide variety of gauges have been used: narrow, to reduce costs under severe topographic conditions; extra wide, used in Russia and Spain, to discourage invasion from neighboring countries; and deliberately different, to inhibit interchange so as to gain a short-term advantage for one region over another. In the 1840s in the United States, different gauges were used by the major trunk lines to discourage shipments coming from the frontier from having an easy choice of Eastern ports.

In North America, railway gauge was standardized after significant investments had been made. Other countries, in particular France, prevented major construction of disparate gauges by establishing standards and routes for private entrepreneurs before construction. Retrofitting systems with new rolling stock and locomotives is an expensive proposition, and rationalization of gauges is often done only on new or upgraded trunk routes, as we have seen with the Japanese standard gauge "bullet" trains and Spain's Train à Grande Vitesse (TGV) (Japan's legacy network is narrow gauge, while Spain's is broad gauge).

The lessons we draw from this experience in railroad standardization are: (1) decisions made on network architecture may have long-lasting repercussions, and their potential future effect on network development therefore should be carefully considered; (2) complete standardization may not be necessary or achievable for reasonably efficient networks to develop—some incompatibilities can be tolerated, and may be economically justified; (3) the role of the state in influencing the choice of standards may vary, without necessarily having negative or positive effects; and (4) the significance of standards choices varies over time—at particular periods, small decisions may have a big effect on network evolution.

Despite disparities in system details from the earliest days, a basic model was set for total control of access to a carrier's technology through standards, regulations, and various de facto and de jure processes. This model for control is further illustrated with the telegraph—a symbiotic technology for the steam railroad, in that each needed the other to expand. Eventually, however, control of railway standards made little difference, just as the strategic significance of telegraph patents declined as new communications systems were built.

Telegraph and Data Transmission Standards
The key standard in both early telegraph systems—the Wheatstone-Cooke (1837) and the Morse-Vail (1844)—was the digital codes used for the analog human interface; the dot-dash of Morse code being the most famous. Telegraph patents were not issued for the interface code alone, but for the entire "conception" of an electrical telegraph. Litigation of the Morse patent cases in the United States was effectively used to keep competitors out of the telegraph business for many years after the invention. The telegraph trials were a landmark in U.S. patent law in that the Morse interests claimed rights to all forms of electrical communication based on their patents. Eventually this broad interpretation was denied, and the patents were validated only as to their narrower claims for electromagnetic signal repeaters. This ruling opened up the field for numerous other inventions in telegraphy and telephony (Coe 1993; Lebow 1995; Lindley 1971; Thompson 1947).

It is not unrelated to the development of the model of carrier control through standards that, as we have noted, telegraph operations began as adjuncts to the railway (although sometimes preceding railway development into virgin territory). In some jurisdictions the principal telegraph firms were subsidiaries or working departments of railway companies. On the European continent, telegraph interconnection was an issue even before railway interconnection had become a problem. In 1866 it inspired the founding of the International Telegraph Union (a direct predecessor of today's International Telecommunication Union) to set procedures for European telegraph networks to manually interwork messages between systems and countries (Codding 1986).

Standards were used, as in the railway model, to control traffic, use, and interface to the public telegraph networks (Grodinsky 1957; Thomson 1947). With most "switching" done by human retransmission, telecommunications standards were concerned more with harmonizing tariffs and ensuring accuracy of transmission. The earliest character-set standards were essentially "commercial" codes to compress text, expedite transmission, and enhance security among rival firms.

The mechanisms of interface control became more complex when machine telegraph encoding was introduced toward the end of the nineteenth century with the Baudot Teletype. The Baudot Teletype used a process to synchronize telegraph machines, and increased the data transmission rate by combining circuits. It was also capable of printing a message automatically. To achieve this increase in telegraph functionality, however, it was no longer sufficient that the codes ("software") be followed rigorously by the user. Baudot extended control to the telegraph and mandated the exact mechanism of the equipment so that the machines would interwork. Emile Baudot's model was robust enough to last until the mid-1960s (Clokey 1936; Hamsher 1967).

Metaphors can be powerful beyond their immediate domain. The railroad influenced the telegraph and telephone. Baudot's Teletype, a technical advance over earlier telegraphs, had a direct bearing on the development of the contemporary approach to television, as we discuss below. Breaking a mind-set is a difficult proposition and has many implications for developing standards.

From the end of World War I until fairly recently, special-purpose machine-encrypted Baudot devices, although technically not permitted on public telegraph networks, served a critical function for governments and major firms that controlled their own communications channels. Indeed, such machines directly stimulated the invention of the stored-program digital computer for decryption purposes during World War II (Randell 1982; Hinsley 1993).

The international standards-setting bodies eventually established Baudot's code and his Teletype mechanism as the only accepted interface to public telegraph systems. The rigid telegraph/railroad metaphor of total system control stimulated growth of a universal public-switched system. Eventually, however, the system's inflexibility discouraged the evolution of a more sophisticated system that would have facilitated interfacing to computer systems. An alternative, more productive scenario would have been to develop the Telex system into a modern public electronic mail, store-and-forward network, with computer processing, graphics, and file-transfer capabilities, that could be integrated with the evolving public digital transmission networks. But by the time the carriers had become more flexible about open Telex interfaces, it was too late to recapture this lost market.

The Bell Model

The Bell Telephone model, an extension of the original telegraph patent model, used licensing and franchising to develop and maintain hegemony over the market, but applied these tools in a new way (Smith 1985). Interconnection to the network was set by specific standards and was protected under law by an increasing number of patents. The Bell extension of the railroad/telegraph model, although not very complicated, was so powerful that it underlies any analysis of telecommunications and information technology standards. Successful firms in the computer field followed this model until their computers became cheap enough to be distributed so their innate power of conversion and imitation eventually changed the rules.

Telephone systems all over the world followed the Bell model to a greater degree of financial success than either the railroads or telegraphs had ever realized. This was true whether the telephone system was privately or publicly owned and operated. This pattern prevailed for nearly a century, until the increasing power and diversity of computer-based telecommunications equipment (as well as legal challenges and market pressures) finally forced networks to become more open in the 1970s. In the scenario of greater deregulation of terminal equipment and access, the control model for contemporary digital, stored-program switched networks can neither require nor maintain the older characteristics. We will return to this issue in chapter 5, with an emphasis on how policy gridlock delayed the opening up of the Bell system.

The Digital Computer and the Role of Government Investment

The idea of automatic data processing can trace its origins to the Jacquard loom of the late eighteenth century, which used punchcards as a means to automate weaving. Charles Babbage and Ada Lovelace used the Jacquard punched cards for their models of the analytical engine, which, had they been completed in the 1840s, would have been the first stored-program digital computers (Randell 1982),[9] Herman Hollerith reintroduced Jacquard punched-card technology in the 1890s for tabulating machinery, and the punched card remained the basis for standards control in automatic data processing until the 1970s.

Hollerith's firm eventually became the International Business Machines Company. IBM's patents and trade secrets on efficient manufacture of tab cards gave them an effective monopoly on machine computation until the advent of the small computer in the 1970s. From Hollerith's days onward, the tabulating companies took a page from the railroad and telephone firms' book, and maintained control over the use and modification of their machines via license and rental contracts. Even when the U.S. government needed to modify IBM tab calculators during World War II for cryptographic purposes, IBM insisted on maintaining and controlling the machines under a license contract to the military. IBM's use of the interface standard as a control mechanism remained a powerful tool in preserving its market dominance until quite recently, when marketplace changes and various antimonopoly agencies in the United States and the European Community succeeded in modifying this practice (DeLamarter 1986).

As we discussed in chapter 2, the first stored-program, digital all-electronic computers were built in great secrecy in 1943 by the British post office electrical engineers, and were used to help decode German ciphers during the war. The machines represented a radical shift: from the beginning of the computer revolution, these visionaries incorporated real-time data access, algorithms to "recognize" a proper problem solution, and machine self-diagnosis to detect failing vacuum tubes. They "downloaded" signals directly from the airwaves, searched off-line for German grammatical structures, and then passed the data on for further code breaking (Flowers 1976; Hinsley 1993; Randell 1982).

The evidence shows that all subsequent stored-program electronic digital computers, as well as the entire generation of current digital telephone switches and digital transmission systems, derive directly from these World War II cryptographic devices. Applications based on Turing's mathematical theories of computing (published in 1936), Flowers's contributions to electronic digital switching in the 1930s, and the critical work of other contemporary innovators in electronic computing in the United States (such as Nyquist and Shannon's work on information theory), France, and Germany were crucial to the evolution of modern telecommunications (Fano 1961; Randell 1982; Weaver 1949).

The development of computers and telecommunications immediately after World War II was not planned by any central body intent on creating an information revolution. The process began unpredictably and was partly due to the push provided first by the military and later by the commercial possibilities inherent in the new technologies. Yet despite their promise, these technologies entered the civilian sector quite slowly.

The critical lesson to be drawn from the history of the development of the computer industry is that those who champion the development of a "free-market computer industry" do so only by ignoring the many decades and tens (if not hundreds) of billions of dollars the government has invested in developing these technologies. The success of the U.S. computer industry is not the result of the government maintaining a hands-off policy, but rather to having its fingerprints all over the industry, for example, providing substantial financing of the supportive educational and basic and applied research efforts of universities. As the commercial marketplace for computers has grown over the decades, especially since the end of the cold war, the magnitude and direction of federal defense-related information systems research and development efforts are naturally called into question. But the assumption that such investment is no longer needed should not be accepted without justification.

Although World War II may have served as the initial impetus for the development of automatic computation, continued government financing at an extremely high level made implementation possible within the relatively short time of only about ten to fifteen years. Computing and computers became a business because the U.S. defense community, after some early fits and starts, spent tens of billions of dollars on computing technology during the 1950s for three primary purposes: cryptography, air defense, and nuclear energy and weapons research.

The National Security Agency alone funded at least $10 billion of unclassified R&D specifically for computing and computers between 1950 and 1960. The NSA's accomplishments included: reduction of the price of magnetic tape to the point where it became cost effective to produce audio cassettes in the early 1960s; introduction of solid-state logic devices to replace vacuum tubes; and significant advances in software tools. Control Data (originally called Electronic Research Associates) and part of Univac (now Unisys) were early spin-offs from the NSA's efforts to build a strong and financially secure U.S. computer and components industry. Many of the NSA's efforts at building increasingly larger computing machinery for weapons design devices were shared with the Atomic Energy Commission (Snyder 1980).

In parallel with developing the cryptologic, intelligence, and nuclear establishment, creating a continental air defense was the other major federal thrust into computing technology. Billions of dollars were spent on the Semi-Automatic Ground Environment (SAGE) network, which began as a high-priority project in the summer of 1950 after the Soviets exploded a uranium bomb. SAGE gave us real-time machines, cathode ray tube(CRT) graphics displays, the light pen, relatively inexpensive magnetic core memory, and even the first computer game.[10] SAGE's spin-offs included two of the world's largest computer firms: IBM, a novice in computing but quite established by then in automatic tabulating machinery; and the Digital Equipment Corporation (DEC), a new company that found a niche in logic circuits and later small computers.

SAGE also provided the impetus for the formation of several national laboratories that contributed to significant advances in telecommunications and computing: MIT's Lincoln Laboratory, the MITRE Corporation, and Draper Laboratories. At Draper, assembly language techniques were devised so that programmers did not have to manipulate binary registers directly. The modem was invented at Lincoln Labs, and the first data networks were designed to operate over the existing telephone and telegraph circuits. Prior to SAGE, the U.S. Air Force had financed Jay Forrester's Whirlwind real-time computer research at MIT, giving him the resources to develop magnetic-core random access memory. IBM made core memory a commercial product, helped by the SAGE contract. Core memory may have been the single most significant advance spurring the development of computers until very large scale integration (VLSI) and ultra large scale integration (ULSI) memory and logic arrived some two decades later (Flamm 1987; Flamm 1988). SAGE also promoted the rapid development of continental microwave links. One side benefit was that continentwide Direct Distance Dialing became economically justifiable somewhat earlier than had been planned.

In the next decade, the government's desire to improve computing and communications systems resulted in the addition of the space program: NASA and other aerospace efforts pushed computing technology into timesharing, advanced digital telecommunications, and satellite communications. At the end of the 1960s, we had packet networks (DoD's ARPAnet), graphics workstations, word processors, and finally, the microprocessor.

The next major impetus occurred in the 1970s with a shift to consumer applications. This was spurred particularly by the advent of the Japanese electronics conglomerates, which quickly moved the transistor and other developments in U.S. technology into the commercial marketplace with such earthshaking advances as the digital watch and the pocket calculator, not to mention the "Walkman" and the "Watchman." The digital watch and pocket calculator (as well as the VCR) were products of U.S. R&D, but it was the Japanese who benefited primarily from their commercialization (Staelin 1989). This brings us to present-day concerns about the convergence and control of the pace of innovation of fundamental "dual-use" technologies, which may serve both national security interests and entertainment purposes (Defense Manufacturing Board 1989, Department of Commerce 1990).

The Convergence of Computation and Communication

Assumptions regarding the old integrated networks included fixed overhead, fixed bandwidth, physical analog connections, and hierarchical network architecture. The new public (and private) networks that are being built are increasingly digital, although with mixed analog/digital interfaces. Single-mode fiber makes possible even greater efficiency from resource sharing and intense use of all-digital networks. The new technology brings with it a new set of assumptions, including variable bandwidth allocations with logical (instead of physical) connections and a nonhierarchical network architecture with distributed processing nodes and terminals (Zorpette 1989).

The first commercial digital switch was developed by the French in 1971, before computer techniques could be fully integrated into telephony. Computer concepts were found in the first Bell System Crossbar units designed in the 1930s, which had a semiprogrammable device called a "marker" for storing routing tables. When commercial computers became cost effective in the late 1950s, however, it became economical to use stored-program control (SPC) (i.e., computing) for toll switching in North America. SPC permitted the North American integrated Long Lines network to follow the principles of a hierarchical network, as derived from manual operator practices (Solomon 1991). Also by that time, a fundamental shift in telecommunications had begun: end-to-end digitization. Unfortunately, the underlying trends were not recognized early enough to prevent SPC, relay-based machines to become prevalent in the network. The concept of digitization spurred early work on totally integrated telephony—becoming today's Integrated Services Digital Network (ISDN) grab bag, and the telephone operators' Advanced Intelligent Network (AIN) proposals.

One of the fundamental principles that the early computer interface standards established was to multiplex the bitstreams from different sources to save circuit costs. A collateral principle was to design the signaling protocol to handle a variety of noisy channels. The model of

Table 3.1
Telecommunications shifts

Technology
- Virtual end-to-end, digital process connectivity
- user-to-machine,
- machine-to-machine, and
- user-to-user communications

- Faster and faster computer processors
- intelligent CPE and "smart" receivers
- merged "packet" and virtual circuit switching
- variable bitrate ("asynchronous transfer mode"—ATM)

- Shared process and network control
- between carrier and subscriber

- Line of demarcation blurred

Policy implications
- Increased competition for service offerings
- De facto interconnection of public/private networks
- Standards design depends on process definitions

Result
- Flexible and fungible choices among technologies: carriers, rates, tariff and service arbitrage, and standards
- Decline of network integrity
- Pervasive application of computer practices in telecommunications standards-setting

---

Sources: Anania 1988; Solomon 1989.

multiplexed channels was fundamental, and the concept of synchronous bitstreams, moving in chunks or "packets," came together in another radical development: the packet network.

Table 3.1 summarizes the critical shifts in telecommunications technology and their policy implications. In short, "virtual" links (i.e., network connections defined by digital codes as opposed to physical wiring plans) have transformed connectivity for users and machines. Faster computer processors supporting bursty traffic now power "intelligent" information appliances in the home and office, not to mention in planes, trains, and automobiles. With increased, distributed intelligence, the demarcation of the network boundary and the user's private space is blurred. One policy implication of this shift in technology is that increased competition for service offerings encourages de facto interconnection arrangements for public and private networks, thereby offering users flexible and fungible choices among technologies, carriers, rates, tariffs, and services. Network integrity may be compromised, however, while standards practices must become more dynamic to keep up with the shifting technology—that is, resembling the chaotic free-market world of computer standards, rather than the traditionally more sedate standards development efforts of telecommunications monopolies. At the same time, however, the broad public interest in enabling every household without regard to income to be able to reach emergency services does not change. Similarly, the positive network externality resulting from the fact that a network becomes more valuable to all its members as more people, machines, and services are connected to it does not change, but becomes more difficult to realize than in a world of "Plain Old Telephone Service."

Researchers at the RAND Corporation had outlined the basics of packet switching in 1964, but it was not until 1968 that attempts to build a packet network began in earnest (Cerf 1981). In the United States, the Defense Department's Advanced Research Projects Agency heavily funded and guided the development of the ARPAnet beginning in 1969; and in France, the Institut de Recherche d'Information et d'Automatique (IRIA) began the Cyclades network in 1973. Both were seminal in developing the standards and protocols used today in digital networks of all types, from LANs to ISDNs.

The packet networks defined new network architectures, tariffing principles (based on volume of information flow rather than distance) and radically different network uses. This shift is just beginning to overtake more than a century of thinking in railroad terms for telecommunications. The ARPAnet defined not simply a new set of transmission standards, but an *entire system architecture* (Roberts 1978).

Telephone switches are no longer mechanical devices that rely on establishing physical circuit paths. Telephone authorities, however, still invest in devices optimized for circuit paths—although modern switches themselves are true digital computers—which architecturally (or rather physically, for the architecture is defined by the program) are not different from the computers that switch packets. This change has profound implications for the present setting of telecommunications standards, and system architecture.

The issues of whether and how an open architecture design can be implemented for communications infrastructure is the subject to which we now turn. The transition from analog to digital systems has a critical effect on network architecture and the associated policy choices for network control.

Open Architecture: Diffusion of a New Paradigm

As is often the case in the evolution of complex systems, one simple, seemingly innocuous change may proliferate radical changes throughout the system until, like a slow chain reaction, it finally explodes. Small changes may alter the very substance of a system. One such shift has been superimposing digital techniques on analog network architectures and analog telecommunications channels.

After several decades, the architecture of telecommunications itself may change. Unfortunately, the existing plant of the old systems has so far resisted change; so today we have a mixed bag of heterogeneous systems with hybrid protocols and messy standards. Ultimately, the network shift may have to be reckoned with as a completely new system that does not follow the older rules of political and economic control.

In 1977, the International Standards Organization (OSI) established a subcommittee to develop an architecture for an open systems interconnect reference model, which is

a framework for defining standards for linking heterogeneous computers ... [providing] a basis for connecting open systems for distributed applications processing. The term open denotes the ability of any two systems conforming to the referenced model and the associated standards to connect. (Stallings 1987)

To make the network truly open, the OSI standards attempt to only define these functions, not how the layers are to perform them. If the layer is a black box, then the Turing model applies, because one does not need to know what is going on in the box, only the inputs and outputs. This permits manufacturers to use any technique to make the black box work, including proprietary techniques. The novel quality here, however, is that any black box should be able to connect to any other black box in a layer above or below it. Of course, the layered function need not be a hardware box; it could be computer software embedded either in an application program or a microchip, or some permutation. To return to the railroad analogy, it is as if a train could consist of a steam, diesel, or electric engine equally well, and that each passenger and freight car could be manufactured by different techniques and owned by different companies. In practice, network protocols are not implemented this way because it is not efficient. Instead, larger boxes may be built to encompass several layers, providing the same result.

The transition to an open digital infrastructure will be neither smooth nor equitable. Defining and interpreting standards are among the prerequisites to establishing open networks. As we noted in the discussion of telecommunications technology trends, as telephone networks have evolved toward distributed computer networks, user control—not carrier control—increases. With the user taking or sharing control of routing, bandwidth allocation, and administration, there will be unexpected applications of the public network. The policy questions raised by these changes include: Who pays for the costs of modernization, and how is it to be implemented? Choices on infrastructure must be made.

The efforts over the past decade to respond with open network models to the potentially anticompetitive networking products of large computer firms and monopoly-based carriers imply a reallocation of resources between public and private networks and among users and vendors. Yet although economic models of how telecommunications should or will work are implicit in these standards processes, almost no explicit mention of economic principles can be found in the official OSI and ISDN standards literature.

Interfaces underlie the fundamentals of OSI, as the United States Federal Communications Commission stated in its May 1986 Report and Order on Open Network Architecture (ONA—commonly known as the "Third Computer Inquiry"). Furthermore, basic network capabilities of the "dominant" carriers must be unbundled and made equally available to all users and information service providers at uniform interfaces.[11] The order has been modified several times with some level of implementation under way today. The FCC basically required the affected carriers to work out their own definitions of openness, standards, and network elements with the user community, although final approval resides with the regulatory body. To speed things up, the FCC promulgated a related regulatory concept called "Comparably Efficient Interconnection" (CEI) that treats carrier information service offerings on a case-by-case basis, rather than mandating or specifying new network architecture standards, which the title of the May 1986 report implies.

The FCC had left nearly all technical, economic, and public-interest aspects of ONA, as well as U.S. implementations of ISDN, to the private sector. These include the examination of options, trade-offs, and critical policy decisions. ONA procedures became another complex level of regulatory adjudication. What at first appeared to be a brilliant new concept for redefining the triadic relationship among the common carriers, their competitors, and their customers had become mired in a seemingly endless loop of meetings and negotiations. What was once adjudicated in the tariffs process has simply shifted to the domain of switching software and access accounting. ONA failed to cut the Gordian Knot because it left all the incentives in place for each of the players to position itself for competitive advantage without the existence of true competition to discipline the marketplace. The discipline had to come from adjudication of disputes in the regulatory commissions and the courts—hardly a process likely to lead to innovation and cooperation.

The Inevitability of Open Communications Infrastructure

The nineteenth-century infrastructure paradigms called for "interfaces" that were dimensional in scope—track gauge, Baudot start and stop bits, voltage and current levels, synchronization signals. A minor change means system elements will not work properly. In nature, minor changes may be regressive, and the organism fails to survive or evolve further. But species that do survive appear to overcome entropy; they contain negative entropy, which is "information," as defined in communications theory. Herein lies a method more useful than the railroad analogy for predicting and guiding future information technologies.

The new telecommunications model subsumes dimensionality within layers that imply compatibility, creates typologies that permit the disappearance of species (i.e., specific hardware or application software) without disturbing the system equilibrium, and encourages implementation of new technologies to increase efficiency and diversity (to aid in survivability), rather than to enshrine vested industries. The parallels to ecology are apparent. Although some may object to the comparison of rational engineering design processes for information technologies to the unpredictability of nature, history in fact shows us that rational design choices may have both negative and positive effects on system evolution and adaptation. We prefer this metaphor of organic growth and network development over the mechanistic vision of railroads carrying loads of information from here to there. Communications are not that simple.

Ecologies are described by typologies of living (and nonliving) system components and taxonomies of how they relate to each other; if such typologies are inherently economic or political, then the taxonomies will be ecopolitical. What we have in our current closed systems of standards for information technologies are outdated categories of customer premises equipment, public-switched networks, and enhanced service and information providers. These distinctions are based on historically grounded economic interests, not on the technology of packet-based digital networks. The tension between technology and economics has been growing and has contributed to policy gridlock and a standoff among vested interests. The strengths of the existing industry sectors are considerable. They can hold off integration and interconnection for a long time, but not forever. The logic of the technology conspires to interconnection.

It may strike the reader as ironic that we both call for a bold federal initiative in this direction yet view interconnected and interoperable digital systems as inevitable and impossible for government to effectively control. The two are not paradoxical. The evolving policy structure under which the United States proceeds will have everything to do with how the infrastructure is rebuilt, by whom, and the extent to which the new information infrastructure will meaningfully contribute to national economic growth. Our central argument is that this is not a question of how much the government should or should not "get involved" in this evolving technology of networks of networks, for involvement is inescapable. The "how much" question quickly degenerates into the familiar and fruitless posturing of the industrial policy debate. The question is how.

4

Networks and Productivity

Paul Krugman has observed that "productivity isn't everything, but in the long run it is almost everything. A country's ability to improve its standard of living over time depends almost entirely on its ability to raise its output per worker" (Krugman 1990). The importance of national productivity growth explains the increasing public and business interest in advanced networks. Innovative network technologies, applications, and services affect productivity, economic growth, and the quality of life. Tools for analyzing the economic performance of networked economies are still rudimentary. Nonetheless, corporate strategies and public policies should be informed by the evolving structure of the global information economy. We argue in this chapter that an Open Communications Infrastructure policy framework is best suited for enabling firms and nations to capture the benefit of the new economy.

This chapter analyzes how nations and firms use networked information technologies to improve productivity. It is not inevitable that network investment will lead to economic growth. Information technology investment in lean and agile industrial organizations linked through ad hoc networks is more likely to realize productivity gains than it would in traditional hierarchical firms. These new industrial organizations can flourish in an Open Communications Infrastructure policy environment, raising firm and national productivity growth rates and creating employment and trade opportunities. Education and training are needed to ensure that workers have the skills required to create new wealth and realize social benefits from the Open Communications Infrastructure.

Productivity in the Electronic Agora

Measuring productivity growth and understanding causal relationships between technical change and economic performance is arguably more an art than a science. In this chapter we review some of the evidence on the importance of investment in networked information technology to national and firm productivity. We then discuss the relevance of manufacturing to national economic health and international trade. We also review the emerging evidence on the potential effect of networked education on productivity. There is preliminary evidence to suggest that improved efficiency and higher productivity may be achieved through networked, interactive educational systems.

Telecommunications trade and technology policies are also reviewed in this chapter. The electronic agora, or information marketplace, is not restricted by national boundaries. In fact, it affects Europe, Asia, and indeed the entire world.[12] The transition of the U.S. economy from self-sufficient fortress, impervious to external market conditions, to a player in the global market competition, is summarized.[13] How Europe and Japan are grappling with these challenges through telecommunications, trade, and technology policies is briefly considered for comparison. We suggest that the more rigid industrial structures of both Europe and Japan may impose impediments to realizing the significant gains in productivity now possible for firms and industries through the introduction of networked information technology as part of the Global Information Infrastructure.

The negative transitional effects of productivity growth, which include job loss among employees working at less efficient firms, as well as the resultant social dislocation of closed plants and unemployment, cannot be ignored. On balance, however, we view the technical trends as positive for the economy and society.

We propose that the economic evidence suggests that an Open Communications Infrastructure can eliminate technical and regulatory barriers to network-driven productivity gains. The creation of new industries and new jobs, and the education and training of employees with the skills needed in an information-intensive economy, are principal benefits of a flexible and agile policy approach.

We also argue that the political culture of the United States is comparatively open to the radical changes in established business and government practices required to take advantage of an OCI. Adoption of policies that favor a flexible regulatory structure and reduced barriers to entry can raise standards of living by accelerating the growth in productivity attributable to investment in information and communications systems, supporting investment in networked information technology, advanced manufacturing systems, and redesigned business practices. Accompanied by renewed emphasis on education and training to raise the skill level of the workforce, these policies can stimulate economic growth and raise standards of living in the information economy of the twenty-first century. The Gordian Knot of political gridlock surrounding network technology innovations is described in other chapters. Here we make the case that continued gridlock is not only bad politics, but also bad economics.

Networked Information Technology and Firm Productivity

Productivity growth is defined as the achievement of a greater value of output with the same cost of inputs. Figure 4.1 illustrates that increased productivity correlates with growth in the gross domestic product (GDP), more commonly referred to simply as national economic growth. However, there is a price to be paid for economic growth. Joseph Schumpeter in his studies celebrated the "creative destruction" of capitalism. More useful technologies and more highly valued products over time drive out less desirable products and services (Schumpeter 1941). Jobs of workers performing the less-valued tasks are eliminated. Ultimately, as Drucker stated, "the fabric of society" is transformed. The positive aspect from a macroeconomic viewpoint is that the loss of jobs in affected industries encourages workers to seek employment in more productive undertakings, and redirects capital investment to more profitable enterprises.

Robert Solow's path-breaking studies in the 1950s concluded that technical change was the most significant contributor to economic growth (Solow 1956). The ability of industry and society to facilitate the development and diffusion of technical change is critical in determining national rates of economic growth. The theory is well understood, but

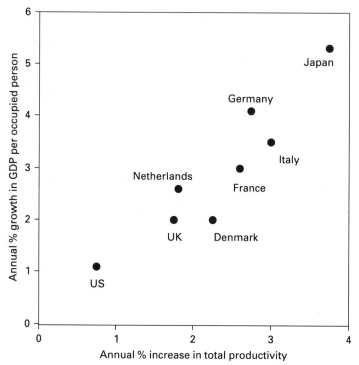

Source: Cornell University, 1991, p. 60.

Figure 4.1
Productivity growth versus growth in GDP

the empirical complexity of assessing national rates of productivity and economic growth is significant (Kuznets 1971). The relative importance for productivity growth of investment in labor or capital equipment, for example, is difficult to ascertain. Measuring productivity in information and service industries is especially difficult (Brynjolfsson 1991).

Traditionally, the economic benefits of investment in information technology and services are measured by the revenue gains attributable to this investment. Increasingly powerful computers and the use of advanced networks might be expected to improve firm productivity and yield a higher rate of national economic growth. Table 4.1 shows the growing intensity of telecommunications investments across business sectors. Coupled with the expansion of the computer industry, there should be evidence of an increased rate of productivity growth. To date, this has

Table 4.1
Telecommunications consumption as a percent of consuming industry output (in millions of 1991 dollars)

	Telecom intensity	
	1965	1987
Telecommunications	0.70	1.80
Finance and insurance	0.81	1.55
Wholesale and retail trade	0.51	1.02
Business services	0.75	0.99
Transportation and warehousing	0.38	0.93
Fabricated metals	0.23	0.90
Personnel and miscellaneous services	0.32	0.83
Electric and electronic equipment	0.42	0.83
Rubber and plastic	0.22	0.78
Stone clay and glass	0.20	0.77
Printing and publishing	0.43	0.73
Radio and TV	1.45	0.59
Instruments	0.44	0.55
Amusements	0.29	0.54
Computer office and nonelectric machinery	0.33	0.54
Primary metals	0.13	0.16
Automotive repair	0.20	0.33
Textiles	0.21	0.29
Mining	0.06	0.27
Furniture	0.24	0.27
Construction	0.11	0.26
Chemicals and products	0.18	0.25
Paper and paperboard	0.15	0.23
Lumber and wood products	0.10	0.23
Leather	0.16	0.22
Motor vehicle and miscellaneous	0.15	0.22
Real estate	0.09	0.20
Agriculture, food, and tobacco	0.14	0.17
Crude petroleum mining and refining	0.03	0.15
Utilities	0.07	0.12
U.S. average	0.27	0.62

Source: DRI/McGraw-Hill, "The Future Contribution of Telecommunications to the Five-State Ameritech Region," April 1993, pp. II–11.

not been borne out by most of the data. The contrary-to-common-sense view of economists employing traditional productivity measures has been that investment in computers has not had a measurable impact on productivity.[14]

A study of twenty U.S. industries from 1968 to 1986 found that computerization in fact "tended to reduce productivity growth, and as an investment was less efficient than other capital spending" (Zachary 1991). Excessive faith in the value of information and the use of information technology to waste time playing with spreadsheet graphics and wordsmithing memos are among the culprits identified by other researchers (Loveman 1988). Paul Strassmann (1990) has shown that by conventional measures there is no relationship between computer purchases by businesses and the firms' profitability. Investment in information technology, if it is failing to increase profit levels, is, by conventional economic measures, a waste of financial resources. Figure 4.2 illustrates this point. The assumption is that if firms were becoming more productive, they would also become more profitable. The cautionary lesson

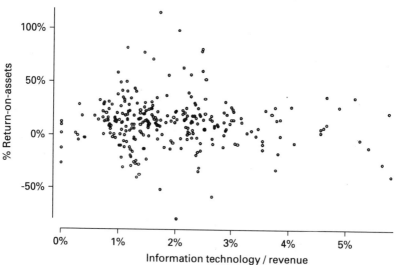

Source: Strassman, 1990.

Figure 4.2
Profitability and computer spending

from these studies is that it is unreasonable to expect all investment in networks and information technologies and services to have positive productivity effects. Substantial sums will be wasted on unproductive information activities and investments (Strassmann 1985). Unpredictable technical progress and regulatory uncertainty increase the risk. The challenge for firms is to reduce the amount of wasted investment without stifling innovation.

For some time, researchers have struggled with the "productivity paradox"—the expectation that information technology and network investment should raise productivity, but the failure to find evidence to support this hypothesis—but only recently has empirical support for a new model of the relationship of networks, information technology, and productivity begun to emerge (Brynjolfsson 1994; Malone and Yates 1987). Recent evidence suggests that the productivity paradox of information systems investment—the apparent failure to increase productivity despite the rapid advance of the underlying technology—no longer exists, if it ever did (Brynjolfsson and Hitt 1993). Part of the problem may have been due to researchers looking at the wrong sort of data, that is, national productivity measures instead of firm-level data sets (Morrison and Berndt 1992). The national data overlooked factors such as improvement in product quality and variety that could increase sales for firms. The use of information systems to redistribute customers across an industry, without proportionately increasing sales, is also missed in the aggregate national data (Brynjolfsson and Hitt 1993).

In addition, the business process redesign or reengineering needed to enable firms to respond to market opportunities through ad hoc networks of corporate partners may only now be paying off (Brynjolfsson and Hitt 1993). It is likely that their lack of network interconnectivity and information system compatibility also contributed to the apparent lack of productivity gains, although the importance of networked information technology versus non-networked information technology remains difficult to ascertain.

Figure 4.3 uses a more recent data set and shows that Brynjolfsson and Hitt found computer spending had a remarkably higher rate of return, and looks very attractive relative to other investments that firms may make to increase their competitiveness. The approximately 80 percent

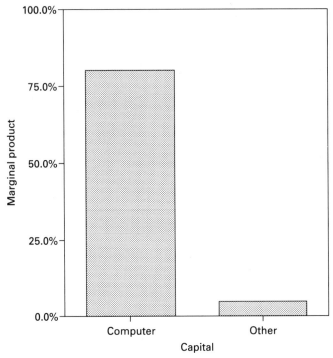

Source: Brynjolfsson and Hitt, 1993.

Figure 4.3
Gross marginal product comparisons

return on investment matches the need for a rapid return imposed by the speedy obsolescence of information technology investment, as technology continues to improve and becomes less expensive (Brynjolfsson and Hitt 1993). In the end the productivity paradox becomes a cautionary tale of the difficulty of measuring immaterial economic activities, the challenge of investing in information technology without squandering resources on incompatible and noninteroperable systems, and the heightened state of competition in the information economy, which requires continual information technology investment just to stand still (i.e., to keep up with competitors and satisfy new user demands and expectations).

As is typical in this field, however, the methodology for measuring information sector productivity that was used by Brynjolfsson and Hitt was called into question by a more recent study. Landauer argued that, even though it was still positive, the productivity growth found by Brynjolfsson and Hitt is an overestimate because of faulty methodology (Landauer 1995). Indeed, in an earlier study Brynjolfsson noted that the lack of evidence of productivity benefits may be the result of mismeasurement, lags between paying the costs and reaping the benefits, redistribution of market share, or mismanagement (1991). All four of these explanations are partially correct.

Close examination of traditional statistical sources indicates that many data sets are ill-suited; and the accepted rules of national economic accounting are inconsistent with the measurement needs of the information economy. For example, education expenditures are traditionally treated as final consumption, rather than investment[15] (Kuznets 1971). In addition, measures of service and product quality—for example, how well a firm's customers are satisfied with its products and services—are not captured in traditional measures of productivity. The simplistic assumption that a more productive firm would be more profitable ignores the limitations on profitability placed by the competitive pressure of other firms' investment strategies and parallel efforts to improve their productivity and, hence, competitiveness.

The benefits of investment in information technology may, as in the case of education, lag behind investment. But with the rapid rate of technological advancement in information technology, firms have only a short period of time to benefit before their information technology investments are superseded by still newer information systems and processes. In the short term, it may be more evident that one firm is gaining market share over the others than that the aggregate level of productivity is increasing. Finally, with the proliferation of information technologies, the range of potentially harmful management decisions and investment strategies increases (Brynjolfsson 1991). The increasing variety of information technology hardware, software, and services suggests that the number of ways for firms to lose money and reduce productivity through ill-considered (or just unlucky) investment decisions is also growing daily.

Information economics deals with understanding, and measuring, these immaterial economic activities. Charles Jonscher, building on the work of Fritz Machlup (1962, 1980) and Marc Porat (1977), found that by 1980 at least half of all economic activity in the United States involved the processing of knowledge or facts rather than the production of physical goods, with the proportion of the former growing all the time.[16] Information processing or "knowledge work" is an important component of many different occupations. Whether one is a factory worker monitoring computer-aided manufacturing systems, an office worker employing personal computers to process information, or a service sector employee making change, information technology is part of the workplace. Reflecting the broad use and critical importance of information and communications technologies to competitiveness in a wide variety of economic sectors, the demand for information and communications technology has also grown as a proportion of demand for all technology. At the same time, demand for white collar or information labor has also grown as a percentage of all labor.

However, the economic effects of "labor-saving" technologies vary. Some workers may lose their jobs and as a result their purchasing power in the process of firms increasing their productivity. This could potentially depress consumer demand. Whether the new wealth created by the new technologies will be sufficient to offset the negative effects of lost wages is not clear. Following a review of the literature on the effects of information technology on business performance, Jonscher states: "Use of technologies which cut costs will, almost by definition, increase the competitiveness of a nation's output on international markets. This effect apart, it would not generally be possible to state unambiguously that the introduction of information technology will increase demand by an amount greater than, equal to, or less than that necessary to offset the fall in labor requirements per unit of output" (Jonscher 1986a). In the decade since Jonscher reported his finding, nothing has changed in regard to the aggregate effect of information technology on labor markets: the picture is still murky.

Nonetheless, it is clear that the nature of the firm and of the organization of markets in a networked economy are themselves affected by network-based innovation. The rate of technical change in information technology is so rapid, and the size of the information or service sector of the economy so large, that productivity increases in this sector can raise national standards of living. In addition, networked-information systems and services have become critical tools for corporations operating across virtually all industries. To analyze these effects more fully, it is necessary to study the practices of business sectors and individual firms. We find that the industrial and firm structure effects of networking vary across industrial sectors.[17] Interpersonal communication need not be technically mediated for industrial networks to be effective (Axelsson and Easton 1992). Lean and agile manufacturers rely heavily on human networks supported by whatever technology is most effective at that time in achieving business goals (Womack, Jones, and Roos 1991). The issue for a firm is whether a particular investment in networks and information technology enables the firm to better pursue its competitive strategy, and whether its competitive strategy is attuned to market conditions and opportunities (Porter 1980).

The potential role of networked-information technology in raising productivity and hence national living standards is increasingly recognized (National Research Council 1994; Gilroy 1993). The term "network industrial organization" is used in Japan to stress the significance of networks in the broader sense to contemporary industrial organizations (Imai 1989). The interorganizational linkages binding Japanese *keiretsu* and Korean *chaebol* support lean and agile manufacturing and inventory management, and technology transfer. Industrial robots and numerically controlled machine tools are monitored and directed through networks. Networked industrial organizations might continue to benefit from the stability and low-cost capital available through horizontal and vertical corporate linkages. Scale economies may still be significant in many markets.

As firms seek to attune their organizations to the altered competitive environment, new patterns of corporate networking and cooperative value-creation processes are emerging (Imai 1989). Alternative corporate structures can be conceptualized in terms of hierarchies and markets. The boundaries of markets and firms are unsettled by information technology, which may affect the relative benefits of hierarchical organizations versus market structures. The need for coordination, cooperation, and information is increasing as corporate hierarchies flatten and intracorporate linkages increase (Malone and Rockhart 1991). "Virtual" ad hoc

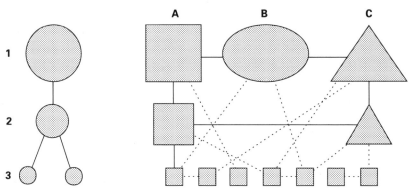

Hierarchical division of labor   Network division of labor

1 = parent firm

2 = medium-sized subsidiary

3 = small firms

A = Firm 1 (parent of medium-sized and small subsidiaries)

B = Firm 2

C = Firm 3 (parent of medium-sized subsidiary)

Source: McKnight, from Ken-ichi Imai, 1989.

Figure 4.4
Hierarchical and network division of labor

or networked corporate joint ventures, research, standards, and product development consortia, as well as government-industry-academic "collaboratories," portend a more profound change in future business and governance structures (B. Axelsson 1992; McKnight 1989; McKnight and Neuman 1995; National Research Council 1993 and 1994). Figure 4.4 illustrates flexible corporate and governance structures that employ an ad hoc division of labor through changing network relationships. The impermanence of the linkages and the speed with which they can be established and torn down are key features of the new environment.

Thomas W. Malone and others argue, however, that hierarchies in general are dysfunctional organizations given the diffusion of information technology; they expect that smaller, more flexible organizations will benefit from this trend (Malone, Yates, and Benjamin 1987; Brynjolfsson, Malone, Gurbaxani, and Kambil 1993). They concluded that "The emergence of electronic markets will reduce the benefits of vertical integration for many organizations. Separate companies—not vertically integrated companies—will perform different steps along the value-added chain, and the exchanges between those independent players will be more efficient" (Malone, Yates, and Benjamin 1989).

Information technology may increase the economic benefits of flexibility, but this flexibility may be achieved in different ways. Information technology, enabling smaller firms to access services and manufacturing resources that in the past were available only to larger, vertically integrated firms, can alter the predominant paradigm of multinational industrial structures and change the economically efficient scale of a business (Brynjolfsson, Malone, Gurbaxani, and Kambil 1993). The falling prices of computer chips and the growing capabilities of fiber and other network technologies reduce telecommunication or networking costs and lower barriers to entry. This leads to the creation of new market opportunities that did not exist before for small firms, and enables businesses to earn a sufficient return on investment within a reasonable time frame to plan for further expansion. Information technology investment can lead to smaller firms becoming more important and contributing a larger share of national and global economic growth (Brynjolfsson, Malone, Gurbaxani, and Kambil 1993). The importance of global brands and global infrastructure for global markets, however, suggests that economies of scale and scope may shift, but will probably not fade away altogether in importance. Evidence that the absolute scale of the largest enterprises is shrinking is lacking in other nations, just as it is in the United States, in spite of the well-known difficulties of particular firms. In an Open Communications Infrastructure, networked-information technologies may nonetheless lower barriers to entry for many small entrepreneurs. Home-office workers equipped with little more than a fax, a modem, a telephone, and a computer can compete with far larger organizations in some markets.

Even more is required to achieve national productivity growth and raise the standard of living. The potential impact of an Open Communications Infrastructure on manufacturing, education, and international trade and competitiveness is detailed in the following sections.

Manufacturing Still Matters

Akio Morita's statement that "American companies have either shifted output to low-wage countries or buy parts and assembled products from countries like Japan that can make quality products at low prices ... the U.S. is abandoning its status as an industrial power" (1986) was perhaps an exaggeration, but the evidence that something was fundamentally amiss in the U.S. economic system is strong. "A service economy cannot neglect manufacturing; if it does, it moves faster and faster toward obsolescence. But only manufacturing in which information is central will thrive in the next century. America is not so much an information-based society as it is a society transfigured by information, the idea racing ahead of each faltering realization. The freedom to be so structured—which means so powerfully flexible—is predicated on the free movement of goods and services which are most densely industrial—including computer manufacturing" (D. Leebaert 1991).

Wages of the average U.S. worker, after increasing significantly in the 1950s and 1960s, stagnated since 1973, because of the decline in U.S. productivity growth rates.[18] Even without real wage increases, American manufacturers' share of world manufactures trade declined, but has recently stabilized, as Dornbush and his colleagues predicted (Dornbush, Krugman, and Chun 1990). Nonetheless, despite some improvement in trade performance and competitiveness in the 1990s, the trade deficit has remained intractable. Some Americans are at increasing risk of being left behind as the skill level required to compete in the global economy is continuously raised. (The ability to employ networks to expand services trade will be addressed in subsequent sections; here we will focus on the impact on manufacturing.) Furthermore, the end of the cold war has enabled increased participation of many nations in Eastern Europe, Asia, the Middle East, and Latin America in the global economy. The benefits of the increased support for democratic institutions and open economies around the world are significant, but they have nonetheless increased the pressure on many American workers now experiencing direct competition with their counterparts in other nations. The North American Free Trade Agreement (NAFTA) and the General Agreement on Tariffs and Trade (GATT) Uruguay Round, which led to the establishment of the World Trade Organization, are a few of the institutional mechanisms that have furthered this trend.

The recognition of the importance of networks and information technology, which do not recognize borders, is an important factor increasing support for an integrated global economy. There is a widely held belief that the United States in particular, and the global economy in general, are in transition to a service, post-industrial, or information economy (Nora and Minc 1980). Services are considered the principal value added by workers in economies dominated by the distribution and processing of information. Hence one might consider the loss of manufacturing jobs an acceptable trade-off in the creation of new service industries. Knowledge workers will not dirty their hands actually producing goods. Some argue that U.S. prowess in services will compensate for the loss of manufacturing.

Critics have pointed out serious flaws in this line of reasoning (Cohen and Zysman 1987). The simplistic picture portraying information services is as flawed as that of manufacturing. Not all service jobs are the same; some pay minimum wage, while others offer service workers much higher incomes. It is important to note that trade in services plays a distinct role because of the peculiar characteristics of services, which often make them difficult to move across borders, and also difficult to measure. As Dornbush, Krugman, and Chun (1990) noted, "The share of services in trade is far smaller than that in U.S. GNP: in 1987, services accounted for only 23 percent of our trade, versus almost 60 percent of GNP."

Manufacturing does indeed still matter; but the role of communications networks and information services should not be underestimated. The rapid pace of electronics and computer technologies has benefited other industries, such as automobile manufacturing, which employs digital technologies in critical components of the manufacturing process, and in the final product. Jonscher (1986b) finds that "the demand for information workers has grown in recent years principally as a result of the increased requirement for information services ... by those parts of the economy concerned with physical production." Research suggests that one should think of a manufacturing firm as a "knowledge-based information processor ... half of labor costs within factories should be labeled costs for information processing, and 60–70% of total labor costs in globally competitive advanced manufacturing industry are really information costs" (Melody 1991).

In spite of the large investment of U.S. manufacturers in information technology, manufacturing productivity growth, which held strong from

1900 to 1970 at an average of 2.3 percent, fell precipitously to almost half that level (only 1.2 percent) over the past two decades. The U.S. share of world manufacturing fell by one third in the 1980s. Evidently, information technology investment alone is not enough to offset other inhibiting factors in U.S. productivity growth. Or as was suggested by other researchers, it may in fact have contributed to the problem, as misguided investment in information technology has displaced more productive investments in other capital equipment, worker training, or new technology.

A firm may expect to control an innovation for a short time, and limited production volume. An alternative strategy emphasizes few major innovations, mass produced in large volumes. Continuous product innovation yields positive externalities including learning-curve benefits, lower long-run average costs, and higher product quality. With discontinuous product innovation, there is no long-run cost curve on which the firm can realize the incremental advantages of continued improvement in production processes (Cohen and Zysman 1987). There may be national preferences for one approach over the other. To mix metaphors, it is argued that American firms tend to search for revolutionary innovations and cash cows, leapfrogging existing technologies and competitors. Japanese and German manufacturers place more faith in incremental product and process refinement, to improve product quality and variety, and hence maintain a competitive advantage (Cohen and Zysman 1987).

American manufacturing is demonstrating strengthened competitiveness through rapid productivity growth. The decline in manufacturing employment in the United States continues, albeit at a slower rate. Regardless of the fluctuations of the business cycle, manufacturing practices affect competitiveness. Manufacturing is not a mechanical exercise, but rather largely an information-processing activity—resulting in the production and sale of goods. High rates of productivity growth may continue, as interoperable network technologies spread through manufacturing and other industrial sectors. It follows that both a competitive manufacturing base and a services base are needed for economic growth, because they are complementary activities. It also follows that an Open Communications Infrastructure could stimulate manufacturing by accelerating innovation and increasing productivity for internal firm coordination and between customers and suppliers. OCI would open markets to new products by eliminating regulatory and interconnection barriers and create new jobs.

The next section addresses the potential role of networked educational services in improving the quality and effectiveness of education and training, and in contributing to national economic growth in an Open Communications Infrastructure policy environment.

Networked Education and Productivity

The potential role of networked education systems in raising productivity is a subject of substantial interest, although there is little empirical research on this issue as yet (Clinton and Gore 1993). But many agree that in "an economy where the only certainty is uncertainty, the one sure source of lasting competitive advantage is knowledge" (Nonaka 1991). Generally, the evidence suggests that stand-alone training systems (such as a teacher in a classroom or CD-ROMs) can be just as effective as networked multimedia systems in educating students and workers. Some scholars, however, argue that networks can play a critical role as part of "a learning infrastructure for all" (Fisher and Melmed 1991). In this section, we discuss the relationship of education to economic growth. We argue that network infrastructure development is needed to meet the information and education needs of future knowledge workers. Investment in telecommunications infrastructure enables but does not ensure productivity growth. We conclude that it is not technology alone that is needed to compete in the global economy, but rather support for the changing needs for human capital formation that is the critical factor.

The benefits of education are typically discussed in abstract terms. Although direct measures of the economic benefits of improved education are problematic, the correlation of growth in GDP with the percentage increase in intelligence quotient (IQ) measures is striking. The Japanese and the Germans are measurably smarter (having higher average IQs) than in the past. At the same time, Japanese and German GDP has increased substantially (McAdams 1991). The rise in IQ can probably not be attributed exclusively to the respective national education systems, nor to infrastructure expenditures. In addition, the significance of IQ scores is much disputed, particularly given the cultural specificity alleged to affect test scores. However, another striking correlation may be noted. The average score of seventeen-year-old high school students on an algebra test, and the percentage of students taking the exam, is substantially higher in all other advanced industrialized nations than it is in the United States. Whereas past forms of industrialization did not demand a high level of numerical literacy on the part of many workers, this has changed in the information economy. Traditional educational goals in the United States are too low, and the methods are not suited to the new needs. While American children go to school for 180 days each year, their Japanese and German counterparts are in class for 220 days studying more math, science, languages, and geography. It should therefore come as no surprise that American children do not know as much.[19] We are not suggesting that students need be locked in sweltering classrooms in July, but rather that networks and information technology can in effect extend the school day and expand the schoolroom to a worldwide network of information resources. An Open Communications Infrastructure would improve education by increasing access of teachers, students of all ages, administrators, and parents to information, educational materials, and each other (McKnight and Neil 1995).

The implications for education and human capital formation summarized here extend beyond the classroom to every individual, home, and business. Using a network to facilitate learning is not restricted to academic research projects and educational institutions. It is characteristic of network applications used by corporations to monitor and improve manufacturing processes. We have already noted that an unfortunate result of the improved operational efficiencies made possible by information technology has been a loss of jobs. There is little evidence that displaced workers find jobs with comparable pay to those they lost. Retraining programs offering counseling, job-training seminars, and education may assist some workers in managing their personal transitions to new jobs. However, what kind of training should be provided—that is, where future jobs will be—is unfortunately extremely difficult to predict, as is whether the retrained workers will have the requisite skills to perform those jobs. Therefore the role of market forces in defining and achieving educational goals should not be underestimated.

Education and employment needs will change in the next century to exploit trade opportunities as they arise in the global information economy. Improved information infrastructure may initiate a "virtuous circle" in which strengthened education utilizing information technology spurs competitiveness gains by U.S.-based workers, which expands trade and spurs further development of the information infrastructure, which spurs improved educational systems, and so on. Although no technology or policy measure can address all aspects of competitiveness, networked multimedia systems will be critical components of the information infrastructure for many firms. Achieving an appropriate balance of competition and cooperation to enable U.S. companies and workers to raise productivity and create innovative networks, products, and services for global markets is a significant challenge. Enabling U.S. educational institutions to nurture a world-class workforce is required to achieve these economic goals. It is not technology alone that is needed to compete in the global economy, but rather support for human capital formation that is the more critical factor. No technical measure is as significant in determining whether capital investment in information technology and networks will prove beneficial to the firm, or the nation, as the level of skill, motivation, and organization of the workforce (Strassman 1990).

The computer (capital equipment) can substitute for labor and offer lower costs for a number of education and business processes. For school districts, advanced information systems can change administrative practices and generate substantial efficiencies, just as computer capital investment has done for business. The role of education in raising productivity and national economic growth in the long term is more difficult to ascertain directly. It is accepted that on average college graduates earn more than high school graduates, who earn more than high school dropouts, and that people with postgraduate training earn more than all of them. The question is whether there is or could be a relationship to educational achievement resulting from increased investment in networked educational technologies and services (Fisher and Melmed 1991; Melmed 1993). Anecdotal evidence exists to support this claim, but more systematic comparative studies are needed.

We suspect that education networking may play an important role in increasing the productivity of education systems, just as it took networks to increase the productivity benefits of investment in computer systems in business. The same may be true for health care (Harris 1993). Networked education and training systems now under development can extend the ability of workers to adapt to changing technology and market conditions. An Open Communications Infrastructure may reduce costs of equipment and services faster, while increasing the range and improving the quality and usefulness of information resources to which students will have access. Firms will hire additional staff to serve new markets or to deal with greater customer demand, stimulated in part by the lower prices and increased product functionality made possible by higher productivity. The knowledge workers of the next century will have an enormous need for information as well as entertainment bytes.[20]

Whether the higher GDP shown in figure 4.5 to be positively correlated with higher IQs is caused by improved educational systems, or by some other factor, is not self-evident. But we believe it is worth trying to improve education through network and information technology investment so that we can become smarter, more productive workers, and more well-rounded individuals. Open Communications Infrastructure policy has a role to play here as well. An OCI should rapidly lower prices and thereby make advanced information technologies affordable to schools and parents sooner. It would also encourage innovation in networked educational services by eliminating barriers to interconnection, and would enhance access to information and communication by students, teachers, parents, administrators, and information and service providers.

We will now shift our focus to the international implications of networks, firm productivity, and national competitiveness. We review in turn the United States, Japan, and Europe.

Trade and Telecommunications

The new rules for domestic policy frameworks and trade in telecommunications services, agreed to by more than seventy nations through the World Trade Organization in 1997, and related technology policy initiatives are discussed below. Nations would like to capture a share of the global economic benefits accruing from changing information and communication systems. The framework for international trade in network

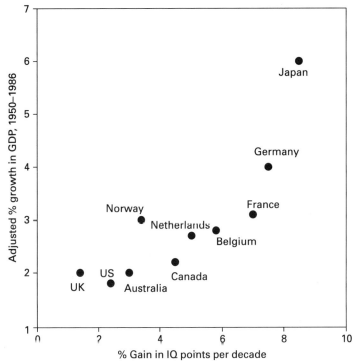

Source: Cornell University, 1991, p. 62.

Figure 4.5
Adjusted percent growth in GDP versus percent gain in IQ, 1950–1986

technologies and services was established for the first time in the Uruguay Round of global trade negotiations concluded in 1993.[21] Recognizing the importance of information markets to global economic growth, the members of GATT chose to extend the rules for trade in goods to include trade in services (Schott 1991). Multilateral agreement on rules for trade in services including telecommunications and intellectual property rights was reached.

Trade in audiovisual materials was removed from the Round, because of strong disagreements between the United States and the European Union (EU) over program production subsidies and broadcast quotas. The dispute on audiovisual materials raised fears that the Uruguay Round would fail, and autonomous trading blocs would limit global trade (Krugman 1990). Others argued that the significance of the GATT negotiations was exaggerated; still more important was preventing the adoption of unsatisfactory rules for international trade (Prestowitz 1991). In the end, the pragmatists won out, and agreement was reached. Although the Round made less progress than was hoped, it nonetheless provided a new benchmark against which further progress in global trade rules can be measured. The ability of the World Trade Organization, which was established by the GATT Uruguay Round agreement to encourage if not enforce free trade, has only begun to be tested by the United States and by other nations, and its ultimate impact is unclear. The WTO agreement will require many nations to open their domestic telecommunications markets to competition, and to establish transparent regulatory processes separating telecommunications operators from regulators. This should accelerate growth and innovation in developing, as well as developed, nations.

The reason that fears and hopes surrounding the establishment of the WTO were raised so high by the international trade negotiations was that public officials recognized that telecommunications and computing were likely to be important sources of innovation in the twenty-first century (OTA 1990). In the international economy, firms and nations use networks to share information and reduce risks—for example, by maintaining constant contact with customers and suppliers (Drucker 1988). Networks can create opportunities for industrial development and international trade.

The United States is far from unique in recognizing the centrality of advanced networks to productivity growth, as the extensive information and communications technology policy programs of MITI and the European Union illustrate. Initiatives for accelerating network technology development and strengthening trade competitiveness in the United States, Europe, and Japan are reviewed below. Indications that hierarchically structured industrial policy programs are declining in their effectiveness are also documented. This is consistent with the findings of Malone, Brynjolfsson, and colleagues in their research on industrial organizations.

The U.S. political economy may be exceptionally able to adapt to the discontinuous technical and regulatory changes required by the emerging global economy. Developing an Open Communications Infrastructure is a prerequisite for sustaining the U.S. competitive advantage in networked markets. On the other hand, it may be too soon to judge which nations will prove most adept at taking advantage of the emerging opportunities while avoiding the numerous obstacles in the way of an Open Communications Infrastructure. On a positive note, significant economic benefits from the globalization of the world economy can flow back to the United States, if it has as open a communications system as possible. We first discuss U.S. telecommunications trade and technology policy, and then turn our attention to Japanese and European network and information technology policy.

U.S. Technology Policy and Telecommunications Trade

The global economy is shifting rapidly, creating and at the same time threatening the competitive advantage of firms and nations. U.S. companies face competitors who enjoy global economies of scale and scope, supportive policy mechanisms, and a well-educated, highly skilled, and motivated workforce. The United States has advantages of its own. We argue that if an agile policy environment can be implemented, an optimistic prognosis for the U.S. future economy can be realized.

In the past, the U.S. economy was essentially self-sufficient, producing domestically 96 percent of everything consumed. That era is over; the percentage of GNP consumed by imports has risen to 14 percent. This is not an especially high percentage by international comparison. What is of more concern is the failure of the United States to sell abroad at an equally increased rate. The U.S. market share of many product categories declined precipitously in the 1980s, even as the U.S. economy apparently recovered from the inflationary fever of the 1970s. U.S. imports of manufactures increased while exports did not grow at equal rates (Dornbush, Krugman, and Chun 1989). More Americans prefer foreign products than foreign consumers prefer American products. In 1989, exports of merchandise amounted to only $252.9 billion, while Americans paid for—or bought on credit—$424.1 billion of imported manufactures (Dornbush, Krugman, and Park 1989). The balance of trade in high-tech products also declined.

Consequently, the American trade deficit increased. The United States shifted from being the largest creditor to the largest debtor nation in the world, reversing sixty years of accumulated commercial development in only four years. Figure 4.6 shows the continuing imbalance of U.S.

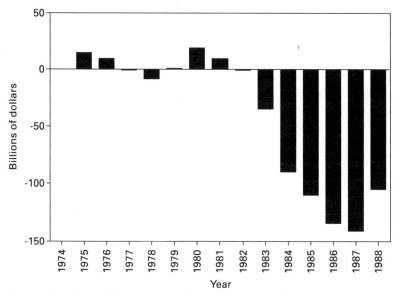

Source: Dornbusch, Krugman, and Chun, 1989.

Figure 4.6
U.S. manufacturing trade balance

manufacturing imports and exports. Federal budget deficits must be financed in part out of foreign borrowing, resulting in a continued outflow of interest payments for many years to come. Some doubt that the United States can continue to run a trade deficit indefinitely; sooner or later a trade surplus will be required to pay back foreign creditors (Krugman 1990). Although U.S. export performance improved and the budget deficit reduced by the mid-1990s, the "twin deficits" in foreign trade and in the federal budget remain obstacles for policymakers.

Direct foreign investment in the United States also increased significantly (Dornbush, Krugman, and Chun 1990). This is not a bad thing; in fact it shows that the United States is considered to have a relatively positive business outlook. Foreign investors appear to be confident that the United States will get a handle on its problems and will be able to provide a good rate of return relative to other global investment opportunities. However, with the decline of communism and the democratization of Eastern Europe, the integration of Western Europe, as well as the privatization and other market-oriented reforms occurring in Latin

America, Asia, and Africa, global investors are enjoying an increasingly broad range of investment options. Foreign direct investment in the United States slowed in the early 1990s. Direct foreign investments in communications worldwide increased significantly between 1981 and 1991, as U.S. and other telecommunications and information technology and service providers scoured the globe in search of investment opportunities (OTA 1993).

Robert Reich argues that we must take care to distinguish between benefits to Americans and benefits to American companies (Reich 1991). Reich correctly points out that what is most important to Americans are "good jobs at good wages," not the nationality of corporate stockholders. To sustain good jobs and good wages, education and training to improve the skills of the workforce are better investments than subsidies to inefficient producers. His unconcern about the fate of U.S. industry, however, is naive. Foreign investment is beneficial, and the global economy should be open, but to believe that the American standard of living will increase at the same time national industry is losing market share may be unduly optimistic. Other nations certainly extend significant efforts to support domestic firms, even as they maneuver in the global marketplace. Japan, Korea, France, and Germany do not base their strategies on the Reichian "What? Me Worry?" formula, depending on the kindness of strangers. Their success at increasing national productivity and economic prosperity suggests that they may know something that Reich and other Americans may have forgotten.

In fact, despite the rhetorical denunciation of mercantilism and protectionism that is part of the American self-image of righteous free-trader, economic development goals have often been pursued. The U.S. record of cooperative and competitive industry-government relationships to strengthen national economic growth is long and complex. U.S. technology policy has long been hidden behind other policy objectives, such as national security or agricultural development. The end of the cold war has focused attention on this previously obscure and widely misunderstood policymaking realm. In the 1990s, information technologies and economic and political changes have made continued neglect of technology policy impossible to sustain. However, confusion about the means and ends of policies for developing communications and transportation infrastructure in the United States remains.

The belief that continued U.S. competitiveness and international political standing is related to the nation's position in information technology markets contributed to some policy breakthroughs in the 1980s. Progress was made even under the Reagan and Bush administrations, which, in their rhetoric, discounted the potential for governments to do anything helpful for industry that would not harm other firms and burden taxpayers. In spite of the rhetorical nuances, there are more similarities than differences between the Reagan/Bush and Clinton administrations' technology policies. A bipartisan consensus on the limits of U.S. technology policy can be discerned, even if unanimous agreement on the importance of technology policy cannot.

We will not provide a detailed history of government and industry initiatives for research and development that seek to develop new markets, but will instead summarize the current practice of technology and industrial policymaking for information-intensive sectors in the United States. Trade and technology policy for information and communications systems in the United States is characterized by a significant degree of ad hoc reaction, incoherence, and ineffectiveness. Competing demands from corporate interests and over twenty federal agencies claiming some jurisdiction over national information and communications policy are joined by the judiciary, numerous congressional committees, state regulatory agencies, and public interest groups, which also voice a vital interest in information infrastructure. The net result is gridlock. To overcome political gridlock, we argue again that development of an Open Communications Infrastructure is needed (Neuman, McKnight, and Solomon 1993). U.S. technology policy, or government initiatives intended to facilitate the development and diffusion of new technologies, is playing a critical role in infrastructure development, just as it is doing in Europe and Asia. Possibly in cooperation with other organizations, governments are supporting research and development of high-capacity, advanced "broadband" digital networks, intelligent components, and high-resolution systems (e.g., digital television).

Technology policy can provide guidance to government, industry, and the public on what to expect from the communications revolution, and what to ask of advanced information infrastructure. For example, the Internet has begun to have an effect on the operations of commercial enterprises, in addition to the educational and governmental organizations that have had access to it all along. Most important, the Internet, built on the technology of statistical sharing (TCP/IP), the economic benefits of positive network externalities, and the policy objective of interoperability, is highly distributed, and maintaining the network is done mostly by volunteers (Bailey and McKnight 1995). The international standards-setting process for telecommunications is undergoing significant reform, but is also creating new threats to U.S. interests. The "bloc voting" potential of a "fortress Europe" may undermine efforts to reach international consensus on standards. It is critical that the FCC and other federal agencies as well as private industry respond in a timely fashion, given that other nations and regions are establishing new institutions and procedures for standardization, testing, and certification to enable their industries to gain early experience with and create a strong market position in, inter alia, personal communications network products and services.

The Internet was developed by a loose coalition of U.S. government agencies, universities, firms, and researchers. It is a preeminent research tool, testing ground, and standards-setter for new educational, scientific, and commercial applications. Multimedia high-resolution digital information and communications services such as videoconferencing, distributed CAD/CAM, and medical imaging are already being tested over broadband networks.

Standards are particularly important for determining conditions of access to networked services, that is, the extent of openness. The economic effects of standards may be enormous. Poor technical choices for standards create high social costs—for example, by limiting the ability of people to communicate freely and by inhibiting innovation (McKnight 1987). There is a constant tension between the goals of ensuring broad access—and hence stable standards—and freezing the development of vital technologies. The European and Japanese research networks described below were intended to influence international networks, for example, through contributions to standards-setting (Drake and McKnight 1988).

A new approach to address this problem was developed by the informal Committee for Open High Resolution Systems (COHRS), in conjunction with a number of standards organizations, industry associations, and federal agencies. Open systems designs for computers and communications have become popular because of the increased connectivity or universality made possible by conforming to open systems requirements. COHRS extended these concepts to digital imaging systems and broadband networks, attempting to facilitate low-cost and potentially universal access to enhanced information services in the future.[22] Although consumer electronics was traditionally thought of as a distinct field from telecommunications and computing, the convergence of technologies underlying these fields is pronounced. American industry had been virtually shut out of this market. However, even if the United States succeeds in establishing a presence in the digital television technology market, for example, technological innovation and economic competition are sure to challenge universal service in the future. The U.S. advanced television research position is strong, but manufacturing capability is problematic in a number of critical technologies (Schreiber 1993).

The transition from the Internet as a dedicated academic research network to a Global Information Infrastructure connecting businesses, government agencies, and residential users worldwide may be problematic. Ensuring interoperability with the public telecommunications network and improving the human interface are among the issues facing the Internet. The bottom-up approach within the Internet community must mesh with telecommunications and computer industry standards and commercial goals. The benefits of an OCI could be realized first in the United States because of the flexibility of its socioeconomic system. The emerging National Information Infrastructure may accelerate technology transfer, shorten product development cycles, and raise productivity. We argue that mutual recognition and reciprocity, which may be adequate principles for relations among peer nations employing comparable regulatory structures, may be insufficient in addressing the new forms of market barriers and discriminatory practices made possible by technical change within asymmetric regulatory frameworks.

European and Japanese telecommunications and information research projects emphasize development of common standards and coordination of precompetitive-technology R&D consortia. Commercial product development is the responsibility of individual firms. Asian and European approaches to both technology and competition policy are different; we do not propose that the United States should copy either of them. We describe some of the Japanese programs in the following section, and point out some of the difficulties caused by failing to implement an Open Communications Infrastructure policy to accompany technology policy programs.

Japanese Network Policy
The growth of Japanese exports since the 1960s illustrates the success of past Japanese industrial policies. The role of *keiretsu* in linking Japanese firms in cross-industry alliances is increasingly recognized; but discussion of Japanese industrial policy practices is still subject to substantial confusion (Okimoto 1989). Despite differences in industry structure, government, and culture, a review of Japanese industrial policies for communications infrastructure development provides an instructive contrast for United States technology policy. The apparent failure of many Japanese industrial policy initiatives for information networks, however, illustrates that the policy mechanisms that worked splendidly to drive the United States away from consumer electronics in the 1970s, and away from dynamic random access memory (DRAM) semiconductor manufacturing in the 1980s, may be less successful in the information markets of the future. A few Japanese programs to promote information network use and development are discussed below.

In Japan, trends toward an "information society" have been widely recognized and accepted by the political and industrial leadership as well as by the public. Coordinated national research and development of electronic technology has been under way since the 1950s. Telephone and television penetration in Japan is on a par with the United States, so that now almost every household has both. Fax penetration is higher in Japan. Due to the complexity of the Japanese written language, personal computers for word processing came into widespread use only in the 1990s. Although Japanese telecommunications firms have not been particularly successful in the U.S. market, consumer electronics has been a significant contributor to the trade gap (Cohen 1991; Cohen 1994). Differences in industrial structure also contribute to the trade imbalance.

In the 1980s and 1990s various national ministries sought to work with industry to develop advanced information networks and services (Levy and Samuels 1989; Okimoto 1989). MITI, as well as the Ministry of Posts and Telecommunications (MPT), the Ministry of Construction, and other organizations have undertaken information infrastructure development programs. Not all of the programs have been successful, but that is to be expected. The intent after all is to reduce, not eliminate, risk.

Government ministries, private industry, and Japanese economic and cultural practices have made significant contributions to the growth of the Japanese economy over the past half-century. MITI, the Japanese Ministry of International Trade and Industry, has taken on mythical status in popular accounts of Japan's rise from postwar ruin to economic superpower. In fact, although MITI has played a prominent role, its practices are not well understood and its rate of success is not as high as imagined. Since deregulation in 1985, the Ministry of Posts and Tele-communications has attempted to maintain its leadership role in Japa-nese telecommunications. The Ministry of Construction and the Science and Technology Agency have funded technical developments as well as information network demonstration projects, which has added to the confusion. And it is always a mistake to overlook the Ministry of Finance when discussing Japanese economic issues. With widespread recognition in Japan of the strategic importance of electronics technologies, com-puting systems, and telecommunications networks, Japanese ministries engage in fierce turf battles and initiate competing projects to claim lead-ership of information network development.[23]

Since the 1970s, telecommunications industry experts expected the next generation of telecommunications to be the development of the integrated service digital network (ISDN). One of the first introductions of commercial ISDN services took place in Japan in 1984. The early introduction was motivated by the desire to give Japanese firms early market experience with ISDN products and services in Japan, to place them in a better position to compete for world markets later on. Despite the large investment already made by NTT, ISDN usage in Japan is much less than had been forecast. ISDN was positioned as an alternative to analog telephony, but commercial users have used ISDN almost exclu-sively for computer-to-computer communication. In addition to the disappointing acceptance of ISDN, CAPTAIN (the Japanese videotext system), and cable, satellite, and high-definition television have not grown as rapidly as expected.

The Hi-Vision (Japan's HDTV) project initiated by NHK in the late 1960s and supported by MPT and MITI was slow to recognize the implications of the digitalization of the media. In Japan, NHK currently broadcasts eight hours per day of HDTV via satellite; but there are very few HD receivers and displays in homes. The consumer market potential of the system appears limited, both because of the expense and only marginal improvement in picture quality. Demand has been far less than originally predicted, and it is likely that their capital investment will not be recovered. The U.S. film and broadcast industry and the FCC rejected 1125/60/2:1 in 1990 in large part because of the problems connected with the continued use of obsolescent interlace scanning technologies. Abandoning interlace requires the Japanese consumer electronics industry to write off billions of dollars in investment in outdated factory equipment. Nonetheless, 1125 is used in some industrial and medical applications both in Japan and the United States where the needs outweigh the technical awkwardness of an analog system in a digital age. Japanese industry also recognizes that more advanced digital progressive scan flat panel displays and open architecture receivers might make the 1125 system more attractive.

Japan's "regional informatization programs" have contributed to the confusion in developing future network strategies. One example of such public-private sector joint projects is the Japanese "Teletopia" large-scale broadband pilot projects. These projects require the cooperation of municipal governments, equipment manufacturers, service providers, and the dominant Japanese telecommunications network operator NTT and encompass entire cities. The ambitious goals for the establishment of Teletopia, Information Network Systems, and Hi-Vision Cities have not been realized (McKnight 1993).

Why have there been so many apparent failures of Japanese industrial policy recently? The critical question of service design was not central to many of the projects, which have focused on hardware design and manufacturing instead. It may be too soon to dismiss Japanese firms from a dominant position in, for example, advanced video system manufacturing and service delivery.

A "new social infrastructure" initiative in telecommunications and information technologies was endorsed in 1993 by the Ministry of International Trade and Industry and the Ministry of Posts and Telecommunications. As part of an economic stimulus package, the program subsidizes computer purchases by universities, research institutes, and schools.[24] Optical fiber networks will be constructed, also with government financial support.[25] The new social infrastructure focuses on the "fiber-to-the-home (FTTH)" project. The Ministry of Posts and Telecommunications is planning a pilot project as a test-bed for a nationwide fiber network that can extend to individual homes. The ministry claimed that during the 1992 U.S. presidential campaign the Clinton-Gore team made a commitment to build a national fiber to the home network in the United States. This of course required Japan to develop a similar project in order not to be left behind.[26]

In March 1993, when the Ministry of Posts and Telecommunications proposed its FTTH policy to the Telecommunication Council, a committee of industry representatives, academics, and other leaders, MPT argued that "the new information infrastructure should be built by the government, not by NTT which is suffering from decreasing profit." NTT has been unable to obtain MPT permission to raise local telephone charges, which are below NTT's costs. NTT challenged the MPT by claiming that NTT alone has the financial resources to invest in fiber-to-the-home, yet it was blocked by the unbalanced telephone tariff structure. The Japanese Ministry of Finance (MOF) joined the debate in support of NTT. In this debate, the Ministry of Finance, which is generally considered to be the most powerful agency in the Japanese national government, is interested in increasing the value of NTT stock.[27] Neither side, however, addressees the question of whether national bondholders, taxpayers, or users of local telephone service should bear the cost of installing future information networks.

Computer communications services, such as NIFTY Serve or PC-VAN, which are similar to Compuserve or Prodigy in the United States, have both enjoyed significant growth, and have apparently benefited from the attention focused on this type of service by the Japanese government.

Despite these new information services, the growth of the Internet in the 1990s has, in fact, left Japan behind. The Japanese Internet was only established in the 1990s to link universities and industry research centers. WIDE, or the Widely Distributed Environment, has its own trunk network of leased lines, but the capacity of its backbone as of spring 1993 was only 192K bits per second—only 1/200th of the American backbone of 45 megabits per second (Aizu 1993). The critical factor explaining the slow growth of networked services in Japan, despite a long-standing industrial policy goal of gaining a leading position in software markets, is the expensive leased-line service costs in Japan. The Internet grew in the United States by linking computers and networks directly via leased lines. In Japan, the prohibitively high cost of leased lines created a bottleneck, slowing the growth of computer-to-computer networks. In 1994, a 1.5 megabit line still cost at least five times more in Japan than the United States.

If tariffs could be realigned, that is, the below-cost prices on local calls raised, and the above-cost prices on leased lines reduced, many Japanese universities, research institutions and small private companies could begin to utilize high-speed networks. Internetworking in Japan could then also grow rapidly, as could the variety of industries that would employ an advanced information infrastructure to offer new products and services to national and global markets.

Japanese development of HDTV and broadband networks is handicapped by the premature standardization of systems that are clearly inadequate to meet the needs of the twenty-first century. Until 1996, Japan's government supported an obsolescent analog HDTV system, which was based on twenty-year-old research. Sony and other Japanese consumer electronics firms' short-term desire to give a boost to the stagnant consumer electronics market persuaded the Japanese government (via NHK, the national broadcaster) to coordinate over a billion dollars of corporate moneys to develop NHK's analog MUSE HDTV system. Billions of dollars in corporate and public funds were spent on technologies even older than that of Europe's antiquated HD-MAC system. It was the lack of corporate vision and weak government leadership in Japan, rather than too much government coordination, that led to the failure of the Japanese industrial policies for HDTV and ISDN. These failures can be attributed to the inflexible, top-down planning processes that were employed, which did not emphasize up-front commercial needs and user requirements, and underestimated the rate of technological change.

The preceding shows that efforts to strengthen Japan's position in the global information economy have not always met with success. The power of entrenched interests, including reliance on the national telephone system and its savings bank to support national economic policy, limited the ability of Japanese policies to adjust to user demand and competitive pressures. Given structural rigidities, an Open Communications Infrastructure policy model would be difficult to implement in Japan. As we write, restructuring of Japanese telecommunications is hoped to help accelerate economic growth after a period of stagnation in the early 1990s. The level of spending in targeted information technology areas indicates the importance the Japanese leadership attaches to developing information markets (Botelho, Cohen, McKnight, and Thurow 1989). See table 4.2. This spending brings only limited benefits to Japanese consumers, because of the barriers to entry and innovation in the ministry-led model of industry development.

U.S. policymakers and business leaders focus on Japanese internal barriers as the culprit for the continuing trade imbalance with Japan. The Japanese government (and many independent economists) emphasize that U.S. domestic policies (for example, a low national savings rate and low standards for public education) are, in the long run, of greater significance than the policies of any foreign government in determining the level of competitiveness and rate of productivity growth of American businesses.

European Telematics Policy

The fall of the Berlin Wall and the "Europe 1992" program for economic integration symbolized the widespread hopes for the elimination of physical and technical barriers to European economic growth. European innovation policy since the early 1980s has focused on telematics, or "télématiques," a term coined by Alain Minc and Simon Nora to describe the convergence of telecommunications and information technologies in their influential report, *The Computerization of Society* (Minc

Table 4.2
Japanese information network projects

Program	Focus	Total cost in US$ (millions)
Teletopia	Broadband networks	10,700
Information Network System	Telecommunication services	8,600
Hi-Vision	Analog 1125-line HDTV;	2,000
	other communications projects	900

Note: Costs are estimated total expenditures of public funds over the lifetime of the project through 1991—which typically represent less than 50 percent of total project investment, because of matching expenditures by private industry. (See also Robert Cohen, 1990.)
Source: McKnight, 1993, European Commission.

and Nora 1980). The European Union economic development strategy since then has been premised on coordination of precompetitive technology R&D consortia and adoption of information technology standards to forge an integrated market. It is hoped that developing common standards will strengthen European industries competitiveness, by enabling their firms to achieve economies of scale within their expanded home market. Some of the programs have begun to extend to Eastern European nations, in advance of European Union membership. These programs to develop and exploit technical change in information and communication systems for competitive advantage are reviewed below.

The goal of the European Union is to build telematics firms that can operate throughout Europe and enjoy economies of scale and scope, and at the same time compete as equals with firms based in the United States and Japan (Grewlich 1992). Public support for research on broadband networks is typically given in the form of 20 percent to 50 percent subsidies for precompetitive cooperative research and development projects. The projects require industrial participants from two or more member states to work together in the European Union. Academic researchers may also participate along with industry and government laboratories. The projects have a finite life expectancy, typically two to eight years. Commercial product development is the responsibility of individual firms. Firms may choose independently to enter into joint ventures, for example, for equipment and services markets expected to emerge from a universally interconnected European broadband network.

A number of information and network technology programs were launched by the European Union in the 1980s. Many other European research programs design network applications. Technology projects include: providing appropriations for continent-wide environmental monitoring networks; developing medical imaging devices and networks; and advancing automotive technology to include a focus on "intelligent" cars and highways—all of which amounts to developing specialized and very sophisticated information networks and services.

Begun by President Mitterrand as a commercial response to the American Strategic Defense Initiative (Star Wars), the EUREKA project is independent from state control, and is quite lean on the ARPA model, with a minuscule staff facilitating the formation of dozens of industry consortia. Although EUREKA can point to some successes, HDTV cannot be considered among them. Preventing the adoption of a worldwide HDTV standard based on proprietary Japanese technologies (the MUSE standard) was considered a crowning achievement of the European Commission. Directorate General 13, Innovation in Communication and Information Technologies, encouraged development of the HD-MAC standard by European industry through about $1 billion in research and development subsidies. The subsidies did produce a substantial number of commercial HD-MAC products which hardly sold.

Despite its relatively strong record, EUREKA suffers from some of the same defects of the other European programs, which will be discussed below. Allowing European policy for HDTV to be dictated by two dominant manufacturers, without regard for consumer preferences, is typical. In Europe, the consumer electronics firms Thomson (France) and Philips (Netherlands) developed the technologies for the analog HDTV European HD-MAC standard, with no direct government involvement. Under pressure from their consumer electronics industries, European governments subsidized but did not direct the technical work (C. Johnson et al. 1992). It was the shortsightedness of the existing European television manufacturing industries that led them to focus on the same obsolescent analog interlaced techniques that are employed by all current television systems—techniques which were state of the art fifty years earlier, but which were abandoned by the computer industry decades ago as being inadequate for high-resolution displays. It is significant that Europe no longer has any major computer manufacturers to push them toward digital technologies, as happened in the United States. Philips and Thomson would prefer a more gradual transition to a digital future, so that they can have time to recover their sunk investment in interlace equipment manufacturing.

Market liberalization and coordinated research efforts were initiated in the context of the Europe 1992 program (CEC 1990). The Research on Advanced Communication Technologies in Europe (RACE) program coordinated dozens of R&D consortia working on a range of technologies for optical networks. The project was officially "prenormative," but is expected to have a substantial impact on the standards, technology, and services of future European broadband networks. The program has since been succeeded by the Advanced Communications Technologies and Services (ACTS) project, which seeks to extend the fruits of RACE research efforts through field trials to pilot-user communities.

The European Strategic Program for Research in Information Technologies (ESPRIT), is able to point to a few European firms growing in niche semiconductor markets. However, an independent European computer hardware industry no longer exists, as American and Japanese firms dominate the market. Since the decline of European information technology firms continued unabated, despite the billions spent by ESPRIT to subsidize research and development, it is difficult to call the program a smashing success.

The ACTS, ESPRIT, RACE, and EUREKA programs are considered by some to be bureaucratic boondoggles, but that is only part of the picture.[28] Collaborative research and commercial ties across the European Union have fostered the psychological emergence in the minds of businessmen, workers, and consumers alike of a common European market—not an insignificant achievement.[29] A number of projects are considered commercially successful, although of course some fail—the outcome of research is uncertain.[30] European open broadband network-related research projects are listed in table 4.3.[31]

Realizing productivity benefits from the vision of convergent telecommunications and computing technologies has proven far more

Table 4.3
European telematics projects

Program	Focus	Total cost in US$ (millions)
Esprit	Information technology	7,600
RACE	Broadband network R&D	3,500
EUREKA	470 precompetitive joint projects in 8 strategic areas, including communication and information technologies	9,300
Telematic programs	Trans-European networks for cooperation in education and training, libraries, linguistics, administration, transport, health care, and rural services	500

Source: McKnight, 1993.

challenging than anticipated by Euro-optimists. Development of robust European information markets is delayed by restrictive regulations.[32] The European Community's industrial policy initiatives ACTS, ESPRIT, EUREKA, and RACE, have produced few demonstrable economic benefits. Many European information technology companies, including Groupe Bull, Siemens, Philips, and Olivetti, have been losing money and market share. EC research projects traditionally encouraged interaction between researchers and manufacturers to align research to market demand and involve users as contributors to the projects. For example, computer networks for health care would involve hospitals and health authorities to jointly define goals for standards and technology research projects. Extending beyond the point of demonstration, at which most previous programs ended, factories could be built to support joint production. The Commission of the European Community in Brussels considered setting up megaprojects for computer networks, and belatedly embraced the Internet.

Productivity and Open Communications Infrastructure

In addition to the intrinsic value of communication to society, infrastructure development issues take on increased significance as nations brace to face intense global economic competition in the twenty-first century. In chapters 2 and 3 we argued that investment in communications infrastructure is the modern-day equivalent to past investment in road and rail transport systems. In the nineteenth century, nations invested in railroad and telegraph infrastructure to support economic development. In the twentieth century, highway networks were established to transport goods and people, while electrical and telephone grids expanded to spread light, power, information, and commerce nationwide. In its present configuration, only a limited number of destinations may be reached via the Internet, the forerunner of a Global Information Infrastructure. However, it also took some time for the railroad, telegraph, highway, and telephone networks to expand to provide convenient national access (McKnight 1993).

The difficult questions of how to design, finance, and govern communications infrastructure should be answered while keeping in mind the potential long-term benefits of infrastructure development.[33] Infrastructure investment is positively correlated with productivity growth, as shown in figure 4.7. Construction of canals, roads, bridges, railroads, and telecommunications facilities will allow the transportation of goods to market, students to schools, and people to jobs faster and at lower costs. Networking can make it possible to accomplish all of this without doing anything more than exciting electrons in fibers. With higher productivity, GDP tends to increase—as does intelligence.

In Europe and Japan, awareness of the strategic economic significance of information infrastructure guides government programs that were begun in the 1980s. Some have advocated that the United States address these challenges by adopting European or Japanese institutional forms, to create a powerful government ministry along the lines of MITI to guide industry and encourage cooperation. European and Japanese firms are encouraged through subsidies and administrative guidance to cooperate to reduce uncertainty about the risks and costs of generic enabling technologies. We argue that the strategic advantage of the United States, rather than relying on government alone, is to employ an array of flexible, voluntary initiatives that encourage cooperation and competition to reduce risks, share information, and raise productivity. It is not enough to work on critical technologies; research funding will only pay off where shrewd technical choices are made and manufacturing is competitive. European, Japanese, and other Asian industrial policies have a mixed

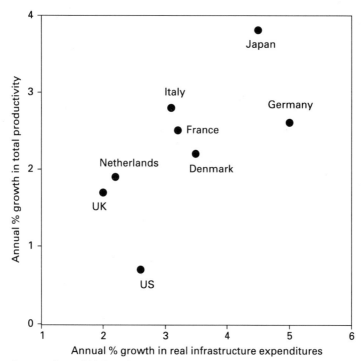

Source: Cornell University, 1991, p. 60.

Figure 4.7
Infrastructure expenditures versus productivity

record, and in any case are not easily transferred to the political culture of the United States.

Public policy should concentrate on improving human capital resources to raise the skill level and health of the workforce. Government's role is to define goals and maintain fair rules of the game, recognizing the positive externalities of investment in education, health, infrastructure, and new technologies, and at times to provide assistance in these areas. Investment in education and information infrastructure development may permit the United States to improve productivity and quality, benefit from flexibility and mobility, and succeed in the increasingly competitive global economy. A policy designed to support the development and use of information networks and services in an Open Communications Infrastructure is more likely to produce new wealth that will benefit all citizens. The continuation of communications policy gridlock clearly is no solution.

But we must also recognize that Japanese and European models will not work here, and American solutions to American problems must be developed. New independent organizations will be best situated to achieve these goals, although one of their main tasks will be to facilitate cooperation and interaction among the existing agencies, corporations, and small businesses, as well as among educators and the public. The logic of the network discussed in chapter 2 suggests that firms and nations, adept at managing the social and economic effects of technical change in global networks, will tend to realize higher productivity and economic growth than nations that attempt to inhibit global economic interchange. This effort must avoid being hindered by the turf-protecting mentality of the existing regulatory structure, or becoming enmeshed in the conflicts inherent in institutional change.

The conclusion drawn from this analysis is that networks can affect educational opportunities and the productivity of individual enterprises, national economies, and ultimately, the global economy. Pursuing open architecture approaches to digital imaging and multimedia communications is consistent with the technology policy goal of encouraging precompetitive cooperation in research projects to develop generic technologies.[34] The challenge is not only to produce quality products as cheaply as possible, but to design a flexible, extensible, and scalable infrastructure to support networked markets. However, network infrastructure without regulatory change is insufficient. A domestic Open Communications Infrastructure may not be mirrored in other nations, thereby promising a continuation of the dynamic of asymmetric regulation and market-access barriers. A government role in monitoring foreign market-access barriers, promoting national trade interests, and protecting domestic workers from foreign predatory practices will continue. Investment in education (human-capital formation) is also needed for networked economies to flourish.

A new U.S. strategy encouraging ad hoc cooperative efforts of government, industry, and academia can have important social benefits. Cooperation through flexible, ad hoc mechanisms might help realize the educational, social, and economic benefits of broadband and personal communications networks. We recall Charles Kindleberger's admonition to embrace change: "The world of foreign trade is one of change. It makes a great difference to the trade of different countries, and to the impact of trade on them, whether they are capable of changing with the world." The realist in us views the economics of investment in information infrastructure and related goods and services and recognizes that there is no guarantee that such investment will increase productivity and trade, or create jobs. The idealist in us hopes that an Open Communications Infrastructure will be implemented and that productivity benefits will be captured, improving the standard of living, quality of life, educational opportunities and overall economic performance of the United States.

We also recognize that the ambiguity of the data on the productivity benefits of networked information technology investment, remains uncomfortably high, as we discussed earlier in this chapter. Although some might consider this uncertainty disconcerting, we find more troubling those who would draw the conclusion that the best recourse is to do nothing. The positive and negative effects of the Industrial Revolution were also disputed at the time of its occurrence. To avoid a new Dickensian *Hard Times* for American workers and society, we believe the best choice is to jump ahead of the curve leading eventually to a more agile policy environment, which can help ensure that the potential benefits of networks for increasing productivity and spreading wealth and knowledge are realized by the United States in the coming millennium.

5

Network Wars: A Pattern Emerges

In this chapter we again address the historical record, but this time our focus is on the policy process rather than on technological and economic trends. We examine the evolution of American regulatory policy permitting interconnection and ultimately competition with the Bell network, on broadcasting, on computers, and on wireless communications systems. Although these might appear at first to be disparate cases dealing with quite different issues, we argue that just the opposite is true.

The history of regulation of wireline, wireless, one-way broadcast and two-way technologies follows a consistent pattern. Each episode starts with what appears to be a relatively minor technological embellishment to an existing system. In each case, the established network interests resist strongly for fear (quite justified as it turns out) that it may weaken their economic position, and in each case the government responds slowly to the new circumstances and bases its decisions on the narrowest technical and bureaucratic criteria. The official public debate pivots on minor, shortsighted issues which are beside the point. The evidence suggests that the American communications policy bureaucracy (with ample and seemingly enthusiastic help from the legislative and judicial branches) is structurally incapable of addressing larger issues of network evolution. The pattern is one of backing slowly into the future with eyes fixed on the immediate past and recognizing the larger issues at stake only later. We argue that these cases illustrate the need for a fresh start, utilizing the best elements of U.S. technology policy to stimulate innovations while avoiding the worst excesses of an encrusted regulatory process.

Telecom Turbulence

First, a brief overview of the central themes of the story. When we think of twentieth-century telecommunications in the United States, we are likely to think of technical and institutional stability—AT&T, the dominant but regulated monopoly, with its far-flung systems, and its Bell Laboratories, a gigantic, successful, world-leading technological juggernaut. But that imagery actually corresponds to the third stage of what is actually a four-stage historical process, a process in which the Gordian Knot has already been cut twice, only to get tangled again because of external events. In the first stage, 1876–1894, the early development of AT&T was protected by the Bell patents, which despite some controversy survived legal challenge and operated as a virtual monopoly. The second stage of telecommunications in the United States, from 1894 to 1913, was a chaotic period of competition and buy-outs as many regions had multiple competing telephone systems. Further attempts to maintain patent hegemony were eroded both by continuing litigation by the government and new inventions not under AT&T's control.[35]

Nevertheless, continuing to aggressively acquire new patents after the original ones expired between 1892 and 1894, and buying up or forcing competition to merge, the Bell interests managed to reestablish ownership of most telephones in the United States—ultimately 80 percent of the national telephone system in the early 1900s. Within a decade, however, the Bell System had dropped to about a 50 percent market share and was under pressure from both the competitive interconnected systems and from the government, which either wanted to break up AT&T under antitrust laws or nationalize it as part of the Post Office (Cummings 1937; Danielian 1939; Horwitz 1989).

Hence, AT&T and its Bell System of franchisees, Long Lines' interconnects, manufacturing subsidiary Western Electric, and core local operating companies were in tenuous legal and financial condition after the turn of the century. Profits were flat, growth was difficult due to internal and external forces, and penetration extended barely beyond the urban business centers and the upper middle class. Long-distance telephony was confined mostly to states in the East. It was rarely used outside of business because of extremely high tariffs, and voice calls could not span the continent. Indeed, about half of AT&T's revenue came from private-line, mostly telegraph circuits, operated for major firms and industries, making AT&T a direct competitor of Western Union, the dominant telegraph carrier, and even of Postal Telegraph, which ran a profitable niche business as the second national telegraph carrier.

In 1906, AT&T's principal banker, J. P. Morgan, brought in Theodore Vail, who had had an extensive career in public utilities, starting with the nascent American Bell Telephone Company in the 1880s, and later moving to South America as a manager of telephone, street railway, and electric public utility properties for Baring Brothers and the Morgan firm (when it took over Baring properties after a major crash). Vail redirected AT&T's external policies toward its customers, including its development of "defensive" technology—in particular, how it responded to public pressures for rate and service regulation. He took over AT&T in an era of increasing antitrust regulation, led by "trust-buster" Teddy Roosevelt, but, moreover, in an era in which the Progressives were calling for nationalization of public utilities and rival financial powers were seeking to diminish AT&T's dominant position in telephony and private-line telegraphy.

Vail ultimately realized that without continued patent protection plus getting the government on AT&T's side, its market share was unsustainable. Rather than fight the government, he came to the conclusion that AT&T would do better to submit to some level of regulation, as a "natural monopoly," which would protect its existing position and erect formidable barriers to entry for competitors, current and potential. Patents, although clearly important from the beginnings of the Bell monopoly, were not enough. Vail's policy was to write a new social contract with the public regarding telecom rates and service, and in exchange the government would be expected to offer further protection from competition. But the process of arriving at this consensus was not straightforward, easy, or rapid. It took about a decade of negotiation, and perhaps another generation before it became received wisdom. Today we are in the midst of renegotiating this Vail "contract," but few have actually realized what the Vail process was or how we inherited this system (Faulhaber 1987).

By 1909, AT&T was in deep trouble, with ownership of less than half of the installed telephones in the United States, and the number of independents having steadily increased from 11,000 companies in 1900 to 13,000 in 1910 (Cummings 1937). Although about 65 percent of the independents did connect to AT&T's Long Lines (Horwitz 1989), the other independents were beginning to interconnect with each other as well, potentially diverting a substantial portion of long-distance revenue, always the most profitable part of the business.

AT&T's initial reaction to competition, despite louder and louder complaints to the Justice Department on grounds of restraint of trade and strong calls for government intervention or takeover, was to refuse to interconnect with independents it considered to be a threat, or to supply them with Western Electric equipment. Pushing matters to the extreme, AT&T had taken stock control of Western Union in 1909, arguing that all telecommunications were a "natural monopoly" and thus should be under the management of one entity.

Neither AT&T's patent strategies nor corporate mergers were enough to hold back either competition or loss of market share. With the increasing success of wireless experiments, new devices being invented to transmit signals long distances, and novel techniques for automated interconnection of circuits—none of which were under AT&T's exclusive control—Vail, always the closet techie with a number of patents to his name, had the foresight to realize that it was just a matter of time before Bell's patent and network position would become untenable.

Contrast Vail's nightmare of being blindsided by change with the lack of vision by the immensely powerful railroad titans of the time. During the same period that Henry Ford was perfecting mass production of automobiles and motor trucks, railroad executives could not imagine that road technology, much less aviation, could topple them from their dominant position in transport. Indeed, as we discussed in chapter 3, the railroads actively promoted the "Good Roads" movement as a means to solving the farm to railhead problem and did not oppose highway interests until the motor truck became a clear and present threat—after World War I.

But more important, railroads wasted precious years and political chits fighting with each other for market share and with their customers over tariffs and service. Ultimately they lost ground as the government extended regulation and antitrust legislation just as new technology emerged to capture critical markets (Goddard 1994; Klein 1987). The railroads had a narrow focus on their immediate environment, believing that: (1) their most important competitors from 1900 to 1914 were the expanding electric street railways and interurbans, siphoning off their short-haul trade; (2) their central political problem was the growing hegemony of labor; and (3) their main economic crisis was deflation, during a period of growing rate regulation based on value of plant. Their reaction was to establish legal and technical obstacles to prevent inter-modal interconnections and to prevent the interurbans from carrying local freight (Hilton 1959). This has the unintended consequence of alienating labor, thereby causing immense inconvenience to their customers with strikes and sabotage. Furthermore, they "gold-plated" their plant to justify rate increases based on the "fair value" of their investment. When technological competition arrived in the form of the motor truck, they had no allies, no cheap local distribution marketing network, and a rigid rate structure under a political process which continued to treat the railroads as whipping boys to please the electorate (Martin 1971). It took another seventy-five years—after innumerable bankruptcies, consolidations, and acres of government reports and studies on the transportation problem—before the railroads could become generally profitable and be considered a strong element of transport infrastructure in the United States, with intermodal technology, flexible rate structures, and productive labor practices (Martin 1982). This would be a startling parallel to the world of telecom today if we continue along the road to gridlock.

Back in the 1910 period, Vail began to understand that, based on his lifetime experience with corrupt governments and fickle customers domestically and overseas, and his earlier experiences fighting Western Union over telephone patents, a new path was needed for AT&T to survive as a privately owned public utility, avoiding the railroad's model of industrial gridlock (Faulhaber 1987; Paine 1921). The winds of change were in the air, but it took real prescience—and perhaps some luck—to sense the oncoming tornado.

In retrospect, Vail's compromise was a brilliant move, making AT&T for the next half-century the largest industrial firm in the United States, with guaranteed profits no matter what direction the economy moved, and dominating the telecommunications industry.

Theodore Vail ushered in the third stage in the evolution of telecom regulation: a truce in terms of the so-called Kingsbury Commitment of 1913. He did this at a time when public utility capitalists and Progressives and their labor allies in the state and federal governments began to dig in their heels for a protracted fight over who should set the agenda for American infrastructure. The ultimate expected outcome was nationalization of all such industries on both sides of the issue.

This stage has never been clearly articulated in the policy debates over telecom monopoly. Kingsbury was neither the government giving up the ghost of nationalization in exchange for a legalized monopoly in telecom, nor was it AT&T winning a fight for a corporate Prometheus, based on technological dominance. Kingsbury is a name attached to a radical compromise to break gridlock, which was hammered out during a prolonged period of vast turmoil in politics, economics, and technology between Vail and an ever-changing group of governmental entities. The year of the commitment was just one point in time during a decade-long watershed.

The famous 1913 letter from Vail's Vice President, N. C. Kingsbury, to U.S. Attorney General James C. McReynolds was a response to protracted negotiations with the federal government beginning about 1910. The letter delineated the parameters for a privately financed and owned monopoly for telephony that would be acceptable to both political parties and sellable to the electorate—a de facto monopoly, if not in law, in exchange for AT&T's best effort to provide universal, affordable service (for the urban population at least), with progressive improvements in service and technical proficiency. Implicit in the Kingsbury Commitment was service to the federal government itself, which was critical in a time of increasing stress, and a commitment which Vail went out of his way to consummate as the international crises mounted with the coming "Great War" (Coon 1939; Danielian 1939; Horwitz 1989).

The long process that led to this novel world has certain similarities to the recent past and perhaps the immediate future. We are suggesting that a similar paradigm shift is necessary to release us from today's gridlock. So, it is worth taking a detailed look at the situation in which AT&T found itself in the immediate pre-World War I period, and at how Vail worked his way through the period to understand that perhaps there are ways around today's frustrating gridlock.

AT&T under Significant Attack, 1910–1913

The Justice Department, upon receiving numerous complaints about AT&T's practices, began a preliminary investigation in 1909 to determine whether the Sherman Act had been violated. AT&T's methods, according to a contemporary report by a Justice investigator, included checking "on the business and earning capacity of each independent exchange, securing a few toll stations in a local city or town, taking a part in the local elections, obtaining a franchise for a complete new telephone service from the local aldermen or council, and then entering into competition with the local exchange or purchasing it...." Competition in long distance from independents, who were connecting their lines separate from AT&T Long Lines operation, was met by "the Bell interests ... by buying here and there independent companies which constituted necessary links in the chain and cutting off connection with independents on either side...."[36]

AT&T's continued abuse of its dominant position, despite being a clear violation of the intent of the generally ignored 1890 Sherman Antitrust Act, combined with a shift in the political mood against big business, encouraged a Republican administration to finally act decisively.

Emboldened by significant victories in 1911 against Standard Oil and the tobacco cartels, and the passage of the 1910 Mann-Elkins Act (which empowered the Interstate Commerce Commission to regulate the rates and services of interstate telephone and telegraph companies for the first time), President William Howard Taft's Attorney General, George W. Wickersham took the initiative. He requested from Kingsbury that AT&T stop further acquisitions and mergers until the Justice Department could complete its investigation. AT&T complied in the summer of 1912, the same summer that witnessed a heated three-way presidential campaign among trust-buster, ex-Republican ex-President Teddy Roosevelt running as a Progressive Bull Moose, Republican President Taft, and Democrat Woodrow Wilson. Antitrust enforcement was a key issue in that campaign, with much criticism of Taft for not having brought criminal complaints against oil and tobacco interests after the Supreme

(a)

Figure 5.1
The peacemakers. Corporate battles often ultimately draw in federal agencies to help make and keep the peace. (a) In 1913 N. C. Kingsbury, an otherwise obscure counsel to AT&T, committed the firm to holding off on continued acquisitions in exchange for peace from the trust-busters of Wilson's Democratic administration before World War I. It became known as the Kingsbury Commitment, for seventy-five years the basis for a private but regulated AT&T as America's phone company. (b) Anne Bingaman, during her term as Clinton's top antitrust chief, played a leading role in reviewing the deals and mergers resulting from the digital revolution. (a) Courtesy of AT&T Archives.

(b)

Figure 5.1 (continued)

Court cases were won by the government to break up the cartels. But Wickersham did not believe in sending "great leaders" to prison, "who have been successful in vast industries,"[37] but instead preferred to negotiate consent decrees whereby the affected industries would admit no wrong but, in some way, would "voluntarily" give up market share (Coon 1939; Cummings 1989).

With unbelievable bad timing, Wickersham submitted his report before the election, concluding that AT&T had not violated the Sherman Act and requesting that the Interstate Commerce Commission (ICC) continue the inquiry under its new powers. The independents, with substantial political backing from the Progressives, small businesses, and voters in small towns and rural areas, felt betrayed and pressured the incoming Wilson administration to do something about AT&T. The same interests have been identified as the major political force behind railroad regulatory legislation both on federal and state levels during the early Progressive period (Horwitz 1989; Kolko 1965).

Wilson's new Attorney General McReynolds continued the negotiations with Kingsbury that Wickersham had started. Both the Democrats and Progressives had called for putting teeth into the Sherman Act with clearer indications of when criminal penalties for violation would apply. Taft had steadfastly opposed such penalties even though Wickersham had waffled on at least one other case, against United Shoe, and recommended criminal sanctions. Finally, with the Democrats in office, and the Clayton Act for antitrust and a bill to establish the Federal Trade Commission clearly moving toward passage (in 1914) (Klein 1987; Thoreilli 1954), Vail could see that it was time for a change in attitude. McReynolds offered AT&T some conciliatory words in lieu of a consent decree—the carrot before the stick would hit hard: "The Administration earnestly desires to cooperate with and to promote all business conducted in harmony with the law." McReynolds wrote to Kingsbury, "Without abating the insistence that the statutes must be obeyed, it will always welcome opportunity to aid in bringing about whatever adjustments are necessary for the re-establishment of lawful conditions without litigation."[38]

Kingsbury took the bait, and replied in the famous letter to McReynolds on 19 December 1913 that AT&T would refrain from further *territorial* expansion through mergers and acquisitions, that it would guarantee to interconnect its service with carriers *outside* of its territory, that it would divest itself of Western Union stock,[39] and move toward a goal of universal penetration in the territory it did hold (via cross-subsidization). From one point of view, as Homer Cummings, the attorney general under Franklin Roosevelt who reopened the 1909 investigation some thirty years later put it, "except for the long-distance provisions, [AT&T] accepted the status quo as the situation under which it would operate in the future, and the government tacitly agreed not to make past conduct a basis for litigation."[40]

Quite true, for in exchange for (limited) universal service, Kingsbury gained an implicit understanding that the government would somehow prevent competitors from working AT&T's territory, instead of AT&T doing the dirty work itself, thereby keeping his executives out of jail—a much neater arrangement. This small provision on long-distance interconnection, however, proved to be a major paradigm shift. For with interconnection, Kingsbury adopted the concept of *open* interconnect (at least to firms of equal standing), of providing the equipment for the independents to connect, and a mechanism for future cross-subsidization of local, hence *universal* service. Moreover, gridlock was postponed, for with these provisions, plus less recriminatory dealings with the government, the opportunity for cooperation among the powers that be was opened. This proved critical during the war years, when AT&T provided advanced communications and R&D to the military for nominal sums, while the railroads had to be nationalized and finally operated by the U.S. Army so that munitions could be transported to the ports during labor strife. Later, after the war when other industries were criticized for war-profiteering, AT&T was complimented by Congress for its public service (Danielian 1939).

Nevertheless, Kingsbury's problems were not at an end, but neither were his opportunities. AT&T's wires, also, were nationalized and turned over to the federal post office at the close of the war, under pressure from the Progressive wing of the Democratic Party, represented by Postmaster Burleson. During the period of postalization, the government compensated the AT&T corporation well, it kept operational control, and in effect raised its rates without opposition since the federal government did not have to abide any state regulatory jurisdiction. Horwitz writes that "AT&T ... got a taste of the benefits of conservative federal oversight"[41] with postalization. The public blamed the rate hike on Burleson, and Congress responded by denationalization a year later, in 1919 (Horwitz 1989). But by then, AT&T had learned very clearly what the Kingsbury Commitment had come to mean: that a friendly government could be helpful in executing your business strategy, and that the government could be made to work for you, not against you—the reverse of the railroads bitter experience in the same period (Martin 1971).

The Republicans won the presidential election in 1920, and with the 1921 Willis-Graham Act, AT&T won the right to further consolidation, this time with the explicit antitrust immunity omitted from Kingsbury, and with the support of the independents, which had not been doing as well financially as the expanding Bell companies. Contrast AT&T's regulatory fortunes with the railroads, which with the contemporaneous 1920 Transportation Act were mandated against their wishes an ill-fated consolidation plan to be prepared by the ICC, and a uniform rate of return across the country, irrespective of regional differences (Latham 1959; Martin 1971). The railroads experienced unabated gridlock for decades, progressively losing market share and unable to take maximum advantage of new technologies until relatively recently. The telephone system, however, was now poised for its greatest growth and technological change ever.

And communications technology continued at a rapid pace with wireless and vacuum tube devices changing the telecom landscape, permitting transcontinental and transoceanic telephony, and opening up new businesses with radio broadcasting. But this was no longer a threat, because AT&T had gotten control of key radio patents during World War I and could now negotiate from a position of strength with its corporate rivals.

Despite the actions of Kingsbury and federal legislation, monopoly protection was not simply a federal prerogative, at least not at first. Instead, an evolving state regulatory regime gradually defined AT&T as a protected common carrier, and then with the establishment of the Federal Communications Commission in 1934, further regulation completely constrained competitors from interstate as well as intrastate competition.

In 1982, in a series of actions similar to those almost seventy years earlier, AT&T agreed with the Reagan Justice Department to halt a new set of antitrust actions, and to divest itself from the local telephone business in order, this time, to free itself from regulation, face competition, and enter into the new markets of advanced electronic communications.

The 1982 "Modified Final Judgment" (MFJ)—actually a modification of some one hundred years of almost continuous antitrust activity, if one includes the original challenges to Bell's patents as part of the continuum—represents the fourth stage of telecommunications evolution and the second time a dramatic attempt has been made to cut a Gordian Knot of legal and political entanglements in order to establish new ground rules. As it turns out, however, the resulting mix of regulated and competitive markets of interconnected electronic networks has remained a chaotic, litigious quagmire of cross-cutting jurisdictions and service definitions. Unlike 1913, 1982 did not write a new social or political contract, but instead brought a new crisis.

An Open Communications Infrastructure regime, based on the evolving digital information architecture of computerized telecom, could be a third and hopefully final attempt to successfully cut the knot. We argue that OCI would open up a fifth stage of open, competitive service provision, this time across *all* electronic media, not just among regulated equals as with Kingsbury. Perhaps, we will then have an *Absolutely* Final Judgement (AFJ) and will no longer have the government continually harassing service providers with either inept regulatory practices or constant threats of antitrust sanctions, so that they can get on with providing effective competitive communications products. Other industries have learned how to compete without the government dictating every rule of the game, and despite some of its unique characteristics, there is no inherent reason why telecom cannot learn as well.

In this survey of telecom turbulence, we review in some detail the third and fourth stages of regulatory evolution, focusing on the repeated failure of partial reforms to address the fundamental technical characteristic of an interoperable network of networks. Eventually, we see that a pattern emerges, from the closed nature of networked infrastructure exemplified by the maturing railroad industry in the 1880s, to the advanced, virtual, digital information infrastructures we see as we enter the next millennium.

Deregulation

The course of deregulation in the United States has been greatly influenced by elements unique to the country's history: the position of telecommunications as a regulated monopoly rather than as a public administration and the dispersion of decision-making authority among federal and state, regulatory, legislative, and judicial branches. Deregulation in telecommunications also coincided with a period of increased skepticism in the United States with respect to all forms of economic regulation, and indeed of the capacity of government to affect the society and economy in a positive direction. In 1970, Alfred Kahn's two-volume textbook, *The Economics of Regulation*, was published which sharply expanded the grounds on which regulation can be justified. Later, as head of the Civil Aeronautics Board, he promoted legislation which dismantled it. (Ironically, Kahn's text cautiously avoids recommending deregulation in the telephone industry.) The Carter administration passed legislation deregulating the trucking industry, emphasizing competition rather than regulation as best for controlling rates, and eventually deregulation reached the railroads with the Staggers Act of 1980, which opened a new golden era of railroading.[42]

Notwithstanding these contemporaneous political trends, the forces impelling the United States to deregulate telecommunications arise largely from the assault on telecommunications perceived natural monopoly from technological changes such as microwave transmission, satellites, computer-controlled switching, fiber optics with unbelievable capacity, and time-sharing and distributed computing. These technologies, and the entrepreneurs and competitors who have championed them, confronted a powerful entrenched foe and a regulatory establishment which came belatedly and almost reluctantly to embrace the competitive structure that had come into being.

Players and Stakes in the Telecom Wars

The Federal Communications Commission

The telephone industry was placed under the regulatory authority of the newly created Federal Communications Commission by the Communications Act of 1934, which consolidated regulatory authority of a number of laws dating back to 1887. Under the FCC, one of F.D.R.'s earliest New Deal initiatives, the Democrats intended to bring AT&T under closer scrutiny than it had been in the previous twenty four years when the ICC had jurisdiction, during which time it had been preoccupied with transport crises—basically ignoring telephony. One of the 1934 act's key provisions was a mandate to study AT&T in detail to determine whether further antimonopoly actions were necessary; this resulted in the Walker

Report of 1938—the first publicly documented indictment of AT&T's practices and procedures (Coon 1939; Danielian 1939; Walker 1939).

The act's purpose is to "make available ... to all the people of the United States a rapid, efficient ... communications service with adequate facilities at reasonable charges" (Paglin 1989). The act created a seven-member commission (later reduced to five members) with broad powers to supervise the telecommunications industry. The commission is authorized to compel interconnection and to approve rates as "just and reasonable." The act obligates carriers to provide service upon reasonable request and forbids "any common carrier to make any unjust or unreasonable discrimination in charges, practices, classifications, regulations, facilities, or services for or in connection with like communication service ... or to give any undue or unreasonable preference to any particular person ...," words that were lifted from previous acts that applied principally to railroad transportation.

As we have noted, since the Supreme Court's seminal 1898 case, *Smyth v. Ames*, public utilities have been regulated on the basis of a maximum permissible "rate of return" on their investments. Thus, the act stipulates that "No carrier shall undertake the construction of a new line or of an extension thereof ... unless and until there shall first have been obtained from the Commission a certificate that the present or future public convenience and necessity require or will require the ... construction and operation of such additional or extended line." This requirement serves to prevent the construction of excess capacity financed by the customer, a clause which stemmed from the 1923 Southwestern Bell Supreme Court decision which basically ruled that regulated public utilities could not pay for future plant in the rate base.

Southwestern Bell was more of a reaction to railroad over-building than AT&T excesses, but it had the same effect. Today it constrains telecommunications firms from using rate-payer income to finance speculative plant, such as fiber-to-the-home, unless they can prove that it is needed for normal growth. Hence it is easier to justify more sophisticated use of copper plant, as technologically obsolete as it may be, than to invest in fiber to the end user, which is not yet proven in the marketplace.

Under the 1934 act, regulated telecommunications carriers must gain regulatory approval both for the construction of new facilities and for their tariffs by the commission and/or state public utility commissions. Filings, of course, may be challenged by competitors or the public; which frequently leads to lengthy proceedings.

Price-cap regulation has become popular more recently. To remove the incentive for over (or mis-) investment inherent in rate-of-return regulation, network providers are permitted to sign essentially a contract with an incentive clause. For example, by promising to maintain or reduce rates by a specified percentage each year, the carrier is permitted to keep any additional profits made possible by increasing operational efficiency at a greater rate. Although not without its critics, price regulation is gaining preference over traditional rate-of-return regulation in a growing number of states. But it is still government regulation with most of the faults of external micromanagement, and still regulation based on plant investment. Price caps are not the radical innovation in telecom provision that its boosters like to think they are, nor are they a flexible panacea for an industry increasingly beset with new technologies, blindsided by different ways of doing things, and faced with rapid obsolescence of old technologies. An OCI would be much more efficient in providing interconnection and absorbing revolutionary changes, but, of course, much less protective of embedded plant and embedded thinking than any form of government regulation, caps, or rate-bases included.

The Judiciary

The judiciary's role in communications arises from two sources. Most important has been the adjudication of antitrust suits brought by the Department of Justice. U.S. antitrust law gives judges wide latitude to impose fines, restraining orders, and structural remedies on corporations convicted of unfair competition. In practice, many suits are settled by a consent decree in which the defendant corporation consents to constraints on its future actions without admitting any past guilt.

Second, decisions by independent regulatory agencies, such as the FCC, can be appealed to the courts for review. While judicial review may not disturb findings of fact, the court may determine that the FCC failed to base its decision on the record before it, or failed to interpret its legal mandate correctly. Numerous key steps in the deregulation of communications in the United States have had their genesis in an appeal of an FCC decision or in a court ruling overturning an FCC decision.

Our Open Communications Infrastructure policy model implies elimination of a great deal of unnecessary regulatory and congressional micromanagement of the information economy. The role of the courts, however, may become even more important in an OCI policy environment. Allegations of antitrust violation must still be investigated and where appropriate, adjudicated. The threat of the potential *criminal* liability, as Vail understood when he accepted the Kingsbury Commitment, should act as incentive for self-regulation of anticompetition behavior. It is presumed that business people do not like to serve time in prison.

The Congress

The Constitution clearly grants Congress the ultimate power to determine national telecommunications policy through legislation, although the power of the Commerce Clause to do this was not settled until *Munn v. Illinois* in 1877 (and some aspects of federal reach into local hegemony still have not been resolved).[43] Yet, by conducting hearings and issuing reports on a series of proposed rewrites and amendments to the Communications Act, Congress over the years has quietly built a public consensus in favor of competition without passing a single bill. The work of Congress on this issue has been largely through the Subcommittees on Communications in both the House and Senate. However, the Congress shows extreme reluctance in lessening its usual oversight of the telecommunications industry, perhaps for no other reason than that telecommunications firms are among the largest donors to congressional campaigns and spend enormous sums lobbying its members.

Several serious attempts have been made to fundamentally reform telecommunications regulation in the United States. Each time the failure to overcome bitter interindustry power struggles, congressional posturing and turf battles, and the technically ignorant, backward demands of self-proclaimed public interest lobbyists have offered inauspicious omens. In spite of the broad public, corporate, and government recognition of the critical need to move beyond the regulatory model of the 1930s on the eve of the twenty-first century, in the moment of truth, all sides prefer to hold on to the past rather than embrace an uncertain future in a cyber-space built on an evolving digital information architecture.

As this book was being written, one more attempt was made to rewrite the 1934 act—perhaps a last chance for the Congress to act before legis-lation itself becomes meaningless in the face of ever-changing technology and the continual erosion of legislative definitions by the courts. The Tele-communications Act of 1996, we noted in chapter 1, is both a remark-able accomplishment and at the same time a disappointing step deeper into gridlock. So many interests attempted to shift the balance in their favor that the only ones truly favored are the lawyers and consultants guaranteed full employment by the need of more firms in more industries to fight for their interests, before the FCC and in Congress. Telecom perhaps has become too complex for legislatures to micromanage; and now that the environment for competition has been set by congressional hearings and reports, rewriting the act may be an idea whose time has passed, given that the FCC already has sufficient authority to remove itself from its own form of micromanagement. It would not hurt, how-ever, if Congress simplified goals and directions by stating clearly that the public interest is best served in competitive provisions and an OCI, and left it at that. The rest could be handled in restraint of trade judicial proceedings.

The Executive

In the 1970s, the executive branch agencies charged with formulating telecommunications policy shifted from a small group in Richard Nixon's White House (Office of Telecommunications Policy) to the National Telecommunications and Information Administration (NTIA) in Jimmy Carter's Department of Commerce. During the past twenty years, these offices were led by individuals with strong procompetitive views which they articulated before various committees of Congress, at the Federal Communications Commission, in proceedings before the Department of Justice and in the courts. These views, however, rarely carried the day in the deliberations of the independent FCC, the judiciary, or the Congress. In the 1990s, the Department of Commerce, through its National Insti-tute of Standards and Technology (NIST), and in particular its Advanced Technology Program, has become a leading vehicle for technology policy, including information and communications systems. Commerce's NTIA has in turn been charged with the responsibility for National Information Infrastructure policy development, as articulated by Vice President Gore right after Bill Clinton took office in 1993.

When the new Republican Congress convened in 1995, both NIST and NTIA (and Commerce itself) were in danger of being greatly weakened; and without sufficient guidance from the president, and no definitive action likely from Congress, the industry faces a more uncertain future in terms of gridlock and government policy than has been seen since 1910.

One possible source of guidance may come from the Justice Department's Antitrust Division. Justice, beginning in the 1909–1913 era (Kingsbury) and continuing through antitrust suits filed in 1949 and 1974, has had an extremely important influence on telecom policy, particularly in dealing with the "gateway" issues. The consent decrees that ended the 1949 and 1974 suits prior to any judicial findings established major restraints and mandated wholesale changes in the industry's structure. The 1996 Communications Act preserved a role for the Antitrust Division in preventing anticompetitive abuses of dominant positions in networked markets.

The 1956 Consent Decree, the "final judgment" for the case brought by Justice against Western Electric in 1949, was a landmark decision (but not a shift in paradigms) in that AT&T agreed to keep out of all businesses not directly related to telecommunications, other than fulfilling requests from the military (such as building atom bombs, a sideline by its subsidiary in Sandia, New Mexico), and came just as the computer revolution was beginning to take place. The 1949 suit, which in part resulted from the original investigation mandated under the 1934 act, would surely have been brought earlier had not World War II intervened, with the government deeming AT&T too critical for the war effort to be hassled over mere monopoly practices (Goulden 1968; Walker 1939). We discuss these implications later. But other than restricting AT&T from expanding into allied information technologies, the 1956 judgment changed less than the status quo ante of Kingsbury did, which at least opened the way for interconnection and was a paradigm shift of the first order.

The 1956 decree was formulated at the height of the cold war. The Republican administration under Eisenhower was anxious to dispose of these cases—which it saw as remnants of New Deal policies—and get on with using American industrial might to contain the Soviet Union. Herbert Brownell, Ike's attorney general, commented in his diary that he had been informed that AT&T was performing crucial work in nuclear weapons and that he felt it was unfair of the government to be suing such a patriotic technological giant when it needed it the most.[44] Memories of World War II were still fresh, and after all, its most famous general was in the White House. Furthermore, Western Electric was also building the first transcontinental microwave links for air defense (discussed below), so Vail's old national security card paid off one more time in keeping the government at bay.

As Steven Coll relates in *The Deal of the Century* (1986), the 1956 judgment left a bitter taste at Justice where some of the old-timers felt they had given away the store to AT&T and were itching to reopen the case. (Some of that bitterness was reflected in the Kennedy administration's opposition to the control of satellite communications by AT&T, which resulted in the establishment of Communications Satellite Corporation (COMSAT), thereby keeping AT&T out of early satellite deployment.) Eventually after a new path was paved by MCI, Justice reopened the case in 1974, which resulted in the *Modified* Final Judgment of 1982, breaking AT&T up into 7 "baby" Bells, and a parent, all of which find themselves now competing with each other. The Justice Department may have felt that AT&T preserved their telecoms market in 1956. But what would have happened to the *information* marketplace if Western Electric had been split then and allowed to compete with IBM in computers, or if the monopoly had been eliminated *before* satellites had become viable?

Yet, the future of antitrust legislation is unclear. It is unlikely that industry rivals will ever again permit a natural monopoly thesis to take hold in telecom. But there are no rules for the future, and every once in awhile strains of the old Bell monopoly sound like an easy way out, especially, for example, when it comes to the contentious issue of who builds the last hundred feet of advanced access plant to the customer. But the technologies for local access now range from broadband cellular radio to satellite, as well as fiber, cable and sophisticated copper, and a contrived natural monopoly for that segment of the digital information architecture will never sound the same as when AT&T controlled the wire pairs all the way into your house up to each and every telephone, and indeed, into the handset right against your ear.

Microwaves: The First Technological Inroad

In 1959, in the Above 890 decision, the FCC overruled the objections of AT&T and determined that private right-of-way firms—such as railroads and pipeline companies—could build their own microwave communications systems, operating at frequencies above 890 megahertz. Many firms claimed that AT&T had been slow to provide them with service along their rights-of-way in remote areas, especially service that could handle their special requirements for signaling and control. AT&T charged that the diversion of spectrum to private users would hamper common-carrier service, and adversely affect national security (the cold war mantra when all other arguments sound hollow), and that the loss of volume would have a negative impact on other rate payers. The FCC, however, downplayed the likely economic impact, and considered the national security argument self-serving; after all, the utilities served national security, too. The Above 890 decision created a market outside the telephone industry for microwave transmission equipment and soon there were numerous suppliers in addition to Western Electric. The utility firms were not allowed to sell excess capacity on their microwave systems, however, without coming under regulation as common carriers, nor were they permitted to connect with the public-switched telephone system, except in emergencies.[45]

AT&T responded to the Above 890 decision with its Telpak rate offering, a bulk discount for wideband or multiple-channel service. To prevent arbitrage, AT&T's Telpak tariffs severely restricted sharing and resale. The FCC in the mid-1960s found portions of these rates to be noncompensatory, and the restrictions on sharing discriminatory. This increased pressure on the commission to relax restrictions further on building private networks, now on satellites as well as with microwave, and to allow interconnection with the public networks.

"Don't Hurt the Network"

When it was clear that fighting new technology on economic diversion grounds would not work, AT&T switched tactics in the 1960s. From the beginning of the patent monopoly in the nineteenth century, telephone companies felt they had to have complete control of their system, much like the railroads had control of the locomotives and cars on their tracks, ostensibly for safety and operating reasons. AT&T too claimed that foreign attachments that were not manufactured by their Western Electric subsidiary could harm the network if not built to their specifications, and could cause operating problems. The option of certifying compliance with Bell System operations and standards, as foreign government-owned telephone administrations have done for years for private branch exchanges and other devices, never seems to have entered the equation, except, of course when the U.S. military needed special connections. But then Western Electric made much of the military equipment, so even those circumstances appeared to be "all in the family."

Before computers forced the issue, two other events—in retrospect almost trivial, indeed silly—opened the way for foreign attachments, and in their own way, for eventual deregulation and finally unbundling of terminal equipment from the core network functions. The first arrived under the unlikely name of "Hush-A-Phone," a plastic device that hooked onto the mouthpiece of a handset to cut down on ambient sounds in noisy locations. AT&T claimed the device harmed the network and the FCC took its word for it. After all, as Horwitz notes, "who knew better than Bell if a device harmed the network?" Hush-A-Phone appealed, and the Court of Appeals, with more common sense than either the commission or the monopoly, not only said it could find no evidence of harm but asked what a monopoly provider with complete control of all manufactured devices it used was doing in also determining which appliances were safe to connect and which were not? Clearly antitrust issues were at stake here (Horwitz 1989). AT&T saw the handwriting on the wall and agreed that mechanical devices were all right, even finally allowing plastic covers on telephone books, which were heretofore forbidden (sometimes we have to wonder whether the people who made these decisions would have been certified sane in any other venue), but that *electrical* connections were still verboten.

Figure 5.2
Technologies of convergence. Sometimes little technologies can make a huge difference. In the early 1930s, a small company started selling a useful device called Hush-a-Phone, which hooked on a receiver to allow one to talk in noisy places. AT&T saw any attachment as an infringement on its territorial mandate and vigorously fought the device as a foreign attachment that would, they argued, jeopardize the network. They were correct, it turns out, that foreign attachments would ultimately jeopardize AT&T's dominant control of the telephone network; but the ultimate culprit turned out to be the silicon chip. Courtesy of AT&T Archives.

Next came the Carterfone case, a device that the FCC ruled in 1967 did not harm the network. It also resulted in adding *value* to the Bell System's service by permitting a radio-telephone to serve sparse oil fields and bring additional revenue into the public-switched telephone network. The Carterfone ruling not only opened the network to any attachment that could show no harm, but set up a crude mechanism for establishing no injury. AT&T now required a fused interface, at some cost, but at least providers could now think of expanding the network beyond AT&T's tariffed services and Western Electric's limited set of appliances. AT&T had still not "gotten it"—that competition in attachments was actually good for increasing their revenues by creating a market for new ideas that they did not think of; AT&T was not omniscient, although they may have been omnipotent. But their competitors did realize that, despite the public relations activities of Bell Labs, all innovations did not emanate from Ma Bell. Indeed, good ideas came from the most unlikely places, and customers for these ideas lurked in firms that the Bell System tended to ignore at their peril, such as the growing data-processing industry which needed specialized services, at prices that AT&T did not want to offer.

Enter MCI

In 1963, William McGowan, founder of Microwave Communications Incorporated (MCI), concluded he could build and operate a microwave network between Chicago and St. Louis, provide new services, and lease capacity to customers for far less than the telephone company was charging for leased lines. He applied for permission to construct a 263-mile microwave system from St. Louis to Chicago. MCI promised to provide a range of services not then available from Bell, including:

• a wide range of analog channel bandwidths
• digital transmission using Time Division Multiplexed (TDM) circuits, with prices charged depending on bit rather than bandwidth
• duplex, simplex, and asymmetric channels (two-way with different bandwidths in each direction of transmission)
• extensive sharing and permissible resale
• part-time (day or night) or full period leases

According to McGowan, cost savings would be due to newer technology, lower overhead, and avoidance of "gold-plated" overdesigned systems. MCI's proposed system would be engineered in part to a tighter standard, but not necessarily lower quality, than that specified by Bell practice: for example, voice circuits were to be 2 kilohertz each rather than 4 kilohertz, and batteries (rather than standby generators) would supply backup power.

AT&T and the regulated telephone companies countered that approval would result in wasteful duplication of facilities and impair their ability to achieve maximum economies of scale. They further contended that price advantages were primarily derived from "cream-skimming" so that approval would eventually destroy geographic rate averaging. In a 1968 report, a presidential task force on communications policy supported

MCI. Finally, after six years, in 1969, the FCC by a four to three vote approved MCI's application to become a "specialized common carrier" (SCC), allowing it to provide leased intercity transmission services.

The FCC's approval of MCI's application triggered a flood of applications from several dozen firms to build additional SCC facilities. In May 1971, the commission issued its Specialized Common Carrier decision, which formulated an open-entry policy for SCCs subject only to technical and financial competence, finding that competition would bring "broadened consumer choices, dispersed responsibility for the supply of communications, economies of specialization, and provision of a regulatory yardstick and an incentive for technical and service innovation by AT&T." AT&T was permitted to compete fully and fairly in private line services and where appropriate "depart from nationwide average pricing for private line services."

Shortly thereafter, the commission decided in its Domestic Communications Satellite Facilities decision to allow open entry in providing leased line services via satellite. In order to give other firms an incentive and head start in competition with AT&T, the FCC ruled that Bell might use satellites only for basic message telephone service (MTS) for three years.

In both these decisions, the FCC took the position that competition in private line services did not threaten AT&T's monopoly in MTS, and therefore it did not pose a significant economic threat to the subsidization of local service by long-distance revenues.

AT&T responded in a number of ways. First, it attempted to limit MCI's ability to interconnect with its facilities and it refused to allow MCI customers to dial local outgoing calls from an MCI line. In 1975, the commission approved a settlement agreement with AT&T so that MCI could set terms for the provision of facilities to AT&T's nascent competitors on an undisputed basis. Second, AT&T introduced its "Hi-Lo" tariff in November 1973, departing from nationwide rate averaging, charging less for high-density lines, and more for low-density ones. The FCC found these tariffs to be unjustified and suspended them in January 1976—shades of short-haul/long-haul tariffs forbidden to the railroads and telecommunications firms in the 1910 Mann-Elkins Act. Third, in order to compete with an all-digital service proposed by a new competitor, Datran Corporation, AT&T accelerated the introduction of its own dataphone digital service (DDS), using its newly developed data under voice (DUV) technology to carry digital data on existing microwave facilities at low marginal costs. These activities led to protracted proceedings before the commission to establish a set of principles for determining the appropriate "costs" for various services.

AT&T argued that as long as new competitive services, such as DDS, were priced to produce sufficient revenue to cover all their "long-run incremental costs" and provide some contribution to the rest of the business, then the rates were "compensatory." This theory conformed to the neoclassical economic argument that prices should reflect marginal costs, but placed an asymmetric burden on monopoly services to support the basic exchange plant. Competitors argued that each different service should contribute a "proportionate" share to the common costs of the business; that is, the common costs should be "fully distributed" over all services. This leads to a higher measure of "costs" for new competitive services, thus making it harder for AT&T to compete. It is, however, much easier to enforce from the point of view of a regulatory agency.

In September 1976, in the aftermath of a tangled debate which had begun twelve years earlier, the commission adopted the modified fully distributed costs principle as a means for determining appropriate rates for competitive services offered by AT&T. The commission was still wrestling with the development of new accounting procedures for AT&T a decade later to make implementation of these principles easier to supervise when the industry was going through an even more traumatic transition with divestiture of the local operating companies and technological change accelerating. One prime example of what technology can do to the most carefully thought-out scenario was the specter of interconnect forced by digital switching.

Execunet

In its Specialized Common Carrier decision, the FCC thought it had restricted the SCCs to offering only leased line services: dedicated channels leased by a particular business customer for its sole use. In 1973, however, with the growing use of Touch-Tone dialing and the availability of new, all-digital computerized switches from sources other than Western Electric, MCI was able to offer a metered private line service called Execunet. An Execunet customer dialed a local number to gain access to the MCI network. After entering an identification code, the customer dials his long-distance call which is routed over the MCI network to the local exchange in the destination city, and then over local telephone company lines to the final destination, thus bypassing AT&T Long Lines. This was a natural evolution of technology that would have been difficult with rotary pulse dialing without modifying AT&T's own switches. The advantages of specialized access were combined with the ubiquity of the switched-public telephone network and were simply not foreseen by Bell technicians as they modernized the switching system for their own advantage.

MCI's emergence as a true intercity telephone carrier was an awkward moment in regulatory history. The FCC, not yet prepared to declare in favor of open competition in basic message telephone service, supported AT&T in its petition that SCC only authorized private line service and not MTS. MCI sued the commission on the grounds that the Communications Act only gave the FCC authority to approve investments in facilities, and not what services could be provided over those facilities (a position similar to that taken by AT&T in introducing its DDS service). In July 1977, the court of appeals ruled that the FCC had never established through hearings that a monopoly in long-distance telephone service was in the "public interest," and that until it had, MCI might continue with its offering. An interesting turn, for the AT&T monopoly was never officially sanctioned as such, it just was accepted as de facto ever since Kingsbury.

The court's decision was a landmark, the first in an era of landmark decisions, because it opened the way for several other companies to quickly begin offering similar services—in effect opening the long-distance telephone markets to competition. The decision marked a change from a presumption in favor of a monopoly to a presumption in favor of competition. It came at a time when full and open competition in customer-premises equipment was supported by the commission, and also when Washington sentiment was in favor of deregulation in general—for airlines and trucking. It coincided with legislative initiatives designed to ratify and extend the preliminary steps toward competition taken by both the commission and the courts.

The Execunet case was important in another way: it gave the lie to the SCCs' claims to be offering "specialized" services not available from AT&T. Despite the promises to offer advanced digital capabilities and other unique services, 90 percent of MCI and Sprint's revenues today come from ordinary voice-grade lines sold on the basis of lower price, although such lines are increasingly used for low-speed data and fax.

The Floodgates Open

Frustrated by the successive rulings from the FCC opening the gates to competition, AT&T turned toward the legislative branch where lobbying power might accomplish what the company had been unable to achieve through the commission or the courts. In 1976, AT&T persuaded fully 56 percent of the Senate and a substantial fraction of the House of Representatives to co-sponsor a bill ingenuously entitled the Consumer Communications Reform Act. The bill would have raised impassable barriers to further entry by the SCCs and also would have permitted AT&T to buy them out—a legislative Kingsbury for the digital age.

Despite intensive lobbying and acceptance of substantial campaign contributions, the U.S. Congress is not easily persuaded to address an issue that puts most voters to sleep, and which is so technical that very few congressional members fully understand what is at stake. This is as true today as it was twenty years ago, and is getting more so as the infrastructure becomes more digital and uses more computerized devices. The actual operation of computers is really very simple, but they perform such complex functions that even people who program them are often not exactly sure what they are doing (Weizenbaum 1976; Hofstadter 1979). With the telephone network as it has evolved, no one person can adequately explain all its functions, interconnections, states, and surely not its disaggregate costs.

In the absence of popular demand for action, the issue was referred to the usual committees. Representative Lionel Van Deerlin, a former broadcaster from San Diego, decided to use the occasion to contemplate a major overhaul of the Communications Act, covering both common-carrier and broadcast issues. In early 1978, after extensive hearings and staff work, he submitted a proposed rewrite of the act which, far from ratifying AT&T's monopoly status, called for increased competition and the use of market forces to replace regulation. Although the bill's provisions for deregulating broadcasting condemned it from the start, a congressional bill favoring competition may have influenced the commission's thinking.

Throughout the early debates over competition, AT&T continued to charge that the other common carriers (OCCs, given that they were no longer "specialized" carriers) were cream-skimming. Not only did they attack the low-cost routes, but also, as providers of basic voice services, they failed to pay their proportionate share of the costs of local service. Accordingly, AT&T's response to the Execunet decision was to introduce a new tariff governing the OCCs' access to local plant: exchange network facilities for interstate access (ENFIA). According to AT&T, since toll subsidized local, the OCCs—which after all were diverting toll traffic from AT&T—enjoyed subsidized rates for local access.

Eventually, with FCC mediation, a negotiated settlement was reached according to which long-distance competitors, depending on their gross revenues, would pay an increasing percentage of AT&T's original ENFIA tariff but never more than 50 percent. AT&T thus established the principle of an access charge in support of local service. This concept will surely be revisited now that the data services such as the Internet have blurred the local access costs once again; only this time AT&T and other carriers may find themselves on both sides of the question of what access entails, both as a potential local access bypass provider and a potential Internet access provider. The access principle under a universal digital information architecture can only be resolved if the concept of OCI takes hold, whereby providers will have to negotiate with each other for interconnection, and those that refuse will only cut themselves out of the greater market.

Successive proceedings after Execunet and ENFIA got even more complex, with each decision opening up competition a bit more. By 1980 the FCC was ruling that full and open competition in all interexchange services was in the public interest, a painfully long time after the technology began to make it all possible about twenty years earlier.

We still have not reached full competitive status for telecom because the local loop is still under a monopoly regime. The technology, however, will surely force that very soon, although some of the local providers have decided to jump the gun and just take a chance on interconnection, dealing with the more ticklish questions of who owns the phone number instead of who pays for each and every element of infrastructure.

Computers and Communication

As we noted in chapter 3, the development of computer technology during and after World War II provided the basis for a major challenge to the regulatory policies established by the 1934 Communications Act. Although demanding that the FCC regulate all forms of telecommunications, the act gave the commission no authority to regulate computation. For a period of nearly twenty years beginning in 1963, the FCC had been struggling to define a means of separating communications and computing.

AT&T's involvement in computing goes back even farther, for much early computer research took place at Bell Laboratories. Indeed, by the mid-1950s when computers finally left the research labs and began to make significant inroads into the business community, the fear that AT&T would use its monopoly telecommunications revenues to finance entry into computers led, in part, to the 1956 antitrust settlement, a suit brought by the Truman administration when computers were just a laboratory toy and the telecom issues were primarily Western Electric's domination of communications equipment.

As we noted, with the 1956 Consent Decree AT&T agreed not to expand into businesses other than communication, or, more precisely, to business not regulated by the FCC. One of the bones of contention was Western Electric's patents for sound motion pictures, which AT&T had used during one period during the 1930s Depression to gain stock control over critical elements of Hollywood production. The FCC's Walker investigation made a particular political football of that one because it put the telephone monopoly in the position of censoring scripts during an era when movies were one of the chief means of propaganda. AT&T quickly removed themselves from that line of fire when top executives discovered their untenable position and claimed the situation had been caused by some rogue lower officials in one of their many subsidiaries— EPRI, Electrical Products Research Incorporated—which sold Western products to non-telephone companies (Danielian 1939; Solomon 1978).

The decree eliminated the line "Sound by Western Electric" from future movies, a line which carried some messy contract clauses about which medium Western Electric-sound movies could be shown in without permission of the company. Imagine the conflicts today if that remained within AT&T purview as we enter the age of multimedia!

Nevertheless, the 1956 Final Judgment was to prove a major impediment to the formulation of sensible policy in later years. By lumping all nontelecommunications equipment with movies and other devices that Western Electric manufactured, computers were confined to stand-alone data processing, which even in the mid-1950s, with the expanding air defense system, they clearly were not. Computers communicate by nature (that is how they work) and it does not matter to the computer if part of its memory is in the next box, next room or halfway across the continent. Given this misperception by the Justice Department (and AT&T), when it came time for the FCC to free certain types of computer communication from regulation, it risked excluding AT&T and Western Electric from any participation or contribution to the field. Indeed, AT&T was careful to call its computer-controlled switches in the early 1960s "stored-program controlled" (SPC) central offices, so as not to sound like it was disobeying the antitrust laws.

We emphasize this point to highlight the way seemingly arcane and obscure regulatory proceedings on communications can have drastic effects on the growth and structure of the American economy. Without AT&T and its communications focus, the computer industry emerged under the dominance of one firm, IBM, with its emphasis on centralized computing—a model radically opposed to the natural evolution of computing as extensions of networks. The lack of an Open Communications Infrastructure policy framework forced the FCC to spend decades attempting to define the undefinable and limit the limitless, as we will now describe.

Time-Sharing Forces the Issue

In the early 1960s, computer "time-sharing," with access via telecom networks was invented, or rather reinvented. Networked time-sharing had been foreseen in part by the continentwide radar air defense computing network built for the air force in the 1950s under contract to

IBM, with Western Electric as the major subcontractor. Indeed, even the very first digital computers built during World War II by the British for code breaking were communicating devices, grabbing encoded messages off the airwaves in real time and then processing the messages. But IBM preferred to follow an off-line model of batch processing for their first machines intended for the commercial market in the 1950s, despite their parallel experience with the air defense designs, because batch processing was an effective extension of the business they knew best, tab card sorters and calculators. So, the first IBM computers in 1954 intended for scientific calculation use punched paper cards for input and output, something that we find very strange today with our display terminals, high-speed laser printers, and networked PCs and workstations.

Time-sharing, introduced in the early 1960s, changed all that. One of the first programs written for these new systems enabled users at a teletypewriter terminal to compose a message and have it printed at another terminal. When Bunker Ramo began offering a computer-based service to stock brokers that allowed them not only to receive price quotes, but also to send messages, Western Union complained to the FCC that Bunker Ramo was engaging in communications common carriage without first having sought a certificate of "public convenience and necessity". Western Union, as well as AT&T, offered similar, but noncomputer-based message services, under the name Telex for the former and TeletypeWriter eXchange (TWX) for the latter.

Bunker Ramo insisted that it was only offering data processing services and should not be regulated. At the same time, Western Union began offering data processing services in conjunction with its advanced record system for the federal Government Services Administration. Competitive data processing firms complained that Western Union had an unfair advantage by virtue of its common-carrier Telex business. These two situations, and others like them, led to the initiation of the First Computer Inquiry by the FCC in 1966 to determine the scope of the commission's authority with respect to online computing, and to establish rules for separating regulated communications from unregulated data processing.

The arguments by the various protagonists in this inquiry fell into the classic predictable patterns we hear whenever any technological change alters the landscape of dominant common carriage. The carriers argued that their commercial provision of data processing would enhance competition in the data processing market. Western Union asserted this would allow it to use its own computers, and hence its infrastructure more efficiently. Opponents noted that dependence on the common carriers for data processing would give the carriers a competitive advantage because they could subsidize low prices for data processing services by drawing on protected monopoly profits from signal carriage. Carriers would also be able to discriminate subtly, even undetectably, against competitors through assignment of inferior quality facilities and refusals to provide service because of alleged violations of tariffs (Oettinger 1977).

These mantras are repeated in virtually every similar case that followed, yet they are quite irrelevant to the question "could computers and computing be regulated as telecom carriage?" The answer the FCC sought was in defining what was computing and what was carriage, and after three attempts in the next two decades, they finally gave up. But the ghosts of those inquiries are still with us as legislators and regulators attempt to define each new twist and turn of the digital information architecture that has evolved.

The commission's 1971 decision, Computer I, attempted to define three categories of service: communications, data processing, and "hybrid" services. Hybrid services would be evaluated according to whether they were primarily communications—in which case they would be regulated —or primarily data processing—in which case they would remain unregulated.

To avoid the possibility of cross-subsidization, common carriers could offer data processing only through "maximally separated" subsidiaries. Any communications services furnished to the subsidiary by the parent would have to comply with tariff (rather than a special contract), in order to minimize the possibility of discrimination, as if computing prices could be tariffed like time and distance services on a communications channel. AT&T was of course barred from engaging in unregulated data processing under the 1956 decree, but Western Union, General Telephone, and other telephone companies were permitted to establish data processing subsidiaries.

Computer I addressed only some of the problems raised by the increasing entanglement of computers and communications. Carterfone, which allowed interconnection of "foreign attachments," was largely motivated by the growing diversity of computers, terminals, and modems, which users wished to connect directly to the public-switched telephone network.

During the 1970s, terminals began to look more and more like computers, as logic boards became integral accessories. When AT&T proposed to offer under tariff its "intelligent" Dataspeed 40/4 terminal cluster, competitors protested that it was a data processing device and thus could not be offered. The commission overruled its own Common Carrier Bureau to approve the 40/4 in 1973.

AT&T had always prohibited the resale of communications capacity to third parties. This was consistent with Bell's notion of a single integrated communications service provided and maintained by one supplier. However, one side effect of the Telpak tariff introduced after the Above 890 decision was to provide a strong economic incentive for a "broker" to buy capacity in bulk under the Telpak tariff and resell it to smaller users, thereby undercutting AT&T. Small firms that would benefit from such brokerage began pressuring the FCC to force AT&T to relax the restrictions on resale and sharing.

Additional pressure for a liberalization of resale came from the Defense Department's Advanced Research Projects Agency ARPAnet (the precursor of the Internet), which used packet switching to communicate between terminals and among computers. Data from one user or machine are interleaved with packets from other users over leased lines and sometimes the telephone network via modems. Sharing circuits this way results in much more efficient use of links. Moreover, in the 1970s, packet switching offered an inexpensive way to translate between different computer protocols, something that today is trivial in the terminal or computer itself because of the vastly increased speeds of processors. In 1973, Packet Communications Incorporated applied for permission to offer a packet communications service. Unlike the SCCs, which constructed their own transmission facilities, Packet Communications intended to simply lease lines from the primary carriers and resell the communications capacity with added value (leading to their designation as "value-added carriers"), very similar to what the ARPAnet was doing for the research community, and in many ways, similar to the way the Internet works today.

This activity led the commission to establish new rules in 1976 for "resale and sharing" of communications. Concerned that the Communications Act required it to regulate all forms of communication, it continued to regulate value-added resellers as common carriers, while relaxing restrictions on entry and pricing. But Computer I's guidelines, with its hybrid categories, quickly proved unworkable in these actions because of such borderline cases. In 1976, Computer II was initiated to seek new definitions. All participants made it clear that a simple definitional solution could not be found. AT&T proposed a service that would have combined packet switching and sophisticated protocol conversion for a wide range of terminals and computers, yet if this was ruled as data processing, it would be forbidden to AT&T under the consent decree.

The commission wanted to make new services such as packet switching subject to open competition. But if it deregulated these services by declaring them to be computing, and thus not subject to the Communications Act, it would in effect prohibit AT&T from offering them. The FCC found itself entangled in a regulatory knot along with AT&T and the emerging data communications industry, from which it could not extricate itself, despite its best efforts. In its April 1980 decision on Computer II, the commission hit on a novel, but equally unworkable approach. It defined two kinds of communication: "basic"—the simple carriage of information from one point to another without any "transformation;" and "enhanced"—which involved code and protocol conversion, storage, and other types of value-added processing. "Enhanced" was still communications, and thus subject to regulation, but the FCC chose a regulatory regime which was, in fact, no regulation at all. This would allow AT&T to offer enhanced services, while allowing the timesharing companies to avoid registering as common carriers.

To ensure that the two largest telephone companies, AT&T and General Telephone and Electric (GTE) could not unfairly compete in the enhanced services marketplace by using revenues from their basic services, both were required to provide any enhanced services through a fully separate subsidiary, as described in the Computer I Inquiry.

In addition, in order to eliminate arguments over whether customer premises equipment was a communication terminal or a data processing device, the FCC ordered that all such appliances (including basic telephone handsets) would have to be sold to the customer by a separate subsidiary rather than be included with tariffed services. This ruling covered not only data terminals but basic telephone handsets as well, thus dropping overnight the entire paradigm for attachments and end-to-end integrated telephone infrastructure. Little by little the walls of telephony were coming down. Computer II also shifted from the commission to the federal courts the responsibility of determining when a service offered by AT&T violated the 1956 decree.

Interestingly, technology was still outracing the regulators. Within two years, the widespread use of PCs and other "intelligent" terminals, obviated the need for value-added networks that did mere protocol conversion. The concepts of basic and enhanced began to become quite muddled; and today with networks of networks such as the Internet and numerous private systems overlaid on the public networks, distinctions such as public and private and basic and enhanced are quite meaningless. The Sturm und Drang of the 1970s and early 1980s, which consumed much lobbying, legal talent, and rate-payer funds, meant nothing in the long run and could have been avoided had the FCC and their mendicants appreciated the steep slope of the technology's curve. Even now we see a failure among policymakers to understand that today's consuming issues are the tomorrow's trivia. Such have become basic and enhanced arguments—mere trivia on the path to a fully integrated digital information architecture.

By the early 1980s, the need to deal with the consequences of open entry in long-distance services became increasing apparent, although the underlying changes coming from the microprocessor had not yet surfaced in the policy arena. It was felt that a system of access fees or other charges had to be established to preserve the cross-subsidies from long-distance or interexchange carriers that kept local access costs low. The competing long-distance carriers were continuing to press Congress for rights to absolutely equivalent technical access to local exchanges still not granted by the FCC. And the American Newspaper Publishers Association wanted assurances that AT&T's newly gained ability to offer enhanced services such as the ill-fated videotex experiments would not lead to AT&T's capturing the market for classified ads through an "electronic yellow pages".

In late 1981, the Senate considered a bill to ratify Computer II by mandating a fully separated affiliate for enhanced services, prohibiting AT&T from providing any "mass media product" (understood to include videotex as well as cable television) over any AT&T-controlled facility, and, even excluding security alarm services. The bill obligated local Bell operating companies to provide interexchange carriers with unbundled and untariffed exchange access equal to what Bell provided to themselves for their own interexchange telecommunications services. It even specified that exchange customers must be offered the option of having all their long-distance calls automatically routed via the long-distance carrier of their choice, that is, MCI rather than AT&T Long Lines.

The bill still seemed threatening to AT&T's rivals, and the House version contained even more safeguards to prevent the gigantic infrastructure built under monopoly from undercutting competition through economies of scale that the new competitors could not even dream of matching. But other events were to take precedence over congressional action, including a final settlement of the 1974 suit to modify the final judgment of the 1956 decree.

The Modified Final Judgment
On 8 January 1982, AT&T and the Department of Justice, in a surprise decision, announced their agreement to a new consent decree that called for AT&T to divest itself within eighteen months of its twenty two local operating companies. AT&T kept control of Bell Labs, Western Electric, and Long Lines, and was freed from all previous court-imposed restrictions on its entry into noncommunications markets. The 1974 suit reopened, and its conclusion thereby modified the 1956 decree, an option the government had reserved at the time. The suit originally called for the divestment of Western Electric and the partial or full separation of Long Lines from the Bell Operating Companies (BOCs).

The 1974 suit charged that AT&T violated antitrust law by refusing to supply interconnection to the other carriers; by adopting engineering and operational procedures that ensured that facilities provided to competitors were of a lower quality than those provided to AT&T's own intercity services; by charging exorbitant rates for local connection; and by engaging in predatory pricing of services faced with competition. AT&T was charged with attempting to monopolize the market by tariffs that prohibited interconnection of attached devices, and then, when those tariffs were overturned by the Carterfone decision, demanding the use of unnecessary "protective connecting arrangements." Finally, Western Electric was alleged to have monopolized the market for switching and transmission equipment through AT&T's control of procurement by the Bell Operating Companies, through the use of Western Electric to evaluate competitive equipment for possible purchase by the BOCs, and through the advance knowledge provided to it by AT&T concerning plans for new services.

The inauguration in January 1981 of Ronald Reagan, a Republican, halted negotiations on a tentative settlement between AT&T and the outgoing Democratic Carter administration. Instead, the case went to trial in March 1981, with the government completing its testimony July 1. In September, the presiding judge denied AT&T's motion for summary dismissal, ruling that the evidence showed that for many years the Bell System committed antitrust violations. The burden clearly would be on AT&T to show that its behavior was compelled by FCC regulation, or was otherwise excusable.

Both tired of continual antitrust litigation—which in all respects was nearly constant since the founding of the company almost a century hence, and fearful of the real possibility that they would lose the case anyway, and might even be subject to the criminal sanctions their executives narrowly missed back in 1913, a small circle of AT&T executives and their board decided in late December 1981 to seek a consent decree involving divestment of all of the BOCs. Part of the motivation was that if they lost in court, they would be forced to divest both the BOCs and Western Electric; and furthermore, Congress would write so many restrictions into the pending legislation that the company would be hamstrung for decades to come. No doubt, the threat of prison terms for top AT&T executives for criminal violations of the antitrust laws—given the evidence that Justice had compiled—motivated them to act as well.

Subsequently, AT&T spun off the BOCs as seven regional companies of roughly equal size. AT&T kept intact Long Lines, Bell Labs, and

Western Electric. Nothing in the decree kept AT&T from offering enhanced services, terminals, or even computers.

The regional BOCs were given responsibility for providing basic local-exchange service, and were initially prohibited from providing interexchange services, customer premises equipment, or enhanced services. At first the BOCs were even more tightly restricted than AT&T had been by the 1956 decree and were to provide nothing beyond tariff-regulated service. Over the past decade the court, under Judge Harold Greene, took responsibility for administration of the decree, and permitted gradual entry of the BOCs into enhanced services via separate subsidiaries and for special services that are considered natural extensions of local-exchange operations; but the reins on the potentially lucrative long-distance services and video programming and distribution were held back. Although the Telecommunications Act of 1996 reformulated the restrictions and moved the venue of battle back to FCC, the transition from regulation to competition is far from over.

Some of the more contentious issues initially raised by divestiture of AT&T, such as preferential buying from Western Electric, which owned the "bell" symbol, and which would coordinate and do research, dio appeared with the sands of time as the seven "Baby Bells" quickly became their own organizations and, indeed, became significant rivals to AT&T and to each other. But other monopoly-barrier questions, not anticipated in 1982, have risen to prominence as technology continues to advance. One of the more intriguing questions is, Who owns your telephone number, now that computer signaling permits full portability? And what is to be done about local competition, now that radio and cable systems may be able to effectively compete with wire-pairs owned by the BOCs? Unfortunately consent decrees, like regulation itself, are very crude ways of dealing with technology that moves faster than litigants can write briefs.

Toward an Open Communications Policy

The evolution of federal policy on interconnection and competition might best be characterized as a series of often desperate attempts to deal with the Pandora's box opened by the Above 890 decision in 1959. Decisions designed to solve a pressing problem of one kind, often created many others, in a pattern of cascading crises. It is the law of unexpected consequences writ large. Seldom was the process able to entertain and evaluate the longer view of technical change. Many of the policies, such as Computer I through Computer III (a 1985 follow-on to the first two inquiries, which attempted to define equitable interconnection and which is almost forgotten in the annals of regulatory wanderings), recognizing the need to take a longer view, did so by attempting to mandate system boundaries and definitions of service types, which were quickly overtaken by the technical capacities and economic incentives for flexibility and interconnection. The task proved to be akin to the labors of Sisyphus: pushing the regulatory rock up hill, only to watch the pressures of technology, economics, and politics pull it back down.

Broadcasting Battles

Broadcasting has always been a battleground for public and private interests, because of its perceived power to affect public opinion and economic and social behavior, its use of scarce resources (spectrum), and its ubiquity. For example, in 1993, the battle for the Russian Parliament made for dramatic live worldwide television viewing. But perhaps the decisive battle had already been fought—and lost. Scores died in the first days of the rebellion as government troops retook full control of Moscow's central broadcast facilities, depriving the rebels of the means to rally public support. This revolution would not be televised by its supporters.

The history of American broadcasting has its own dramatic qualities but is closer in tone to a soap opera brouhaha than the life and death struggles of government overthrows. A balance of public and diverse private interests has relied on regulated monopolies and oligopolies to serve the public and provide open communication in at least one direction (Barnouw 1975). Broadcasters could exercise their rights to free speech, within narrower social boundaries than print media (Pool 1983). Innovation in broadcasting is the outgrowth of an ongoing public-private battle among entrenched institutions. The broadcasting brouhaha began with the birth of broadcasting and continues unabated to this day.

The time has come to abandon the tried and true method of broadcast regulation and adopt instead a new model of open broadcasting, as part of an open communication policy (Neuman, McKnight, and Solomon 1993). But first, we start at the beginning and briefly review the creation of the Radio Corporation of America (RCA), and the nation's broadcast infrastructure.

RCA and the Origins of Broadcasting Infrastructure

Epic political and economic struggles surrounding the development and use of broadcast technologies were resolved for a time with the establishment of a national broadcasting system (Aitken 1985). The origins of U.S. broadcasting infrastructure are found in the public-private patent cartel created to commercialize radio broadcasting technologies at the close of World War I. The cartel was formed to reduce transaction costs and uncertainty due to the technical requirement that transmitters and receivers must be compatible. From the patent pool established by the cartel, the Radio Corporation of America was born.[46] RCA's creation of a national broadcasting infrastructure was a difficult undertaking with an uncertain outcome. Despite its individual and social costs (see the sad tale of Edwin Armstrong below), RCA did succeed in creating a public-private partnership to minimize uncertainty and increase the confidence of manufacturers, broadcasters, and consumers in the stability of the technical and organizational infrastructure for commercial radio broadcasting (Barnouw 1966; Waldrop 1938).

During the Versailles Peace Conference of 1919, President Woodrow Wilson and his advisors realized that three types of international infrastructure would be critical in postwar trade: energy, the merchant marine, and the new technology of radio communications. All three were linked in some manner, especially the latter two. Radio and undersea telegraphy were dominated by the British at that time.

Assistant Secretary of the Navy Franklin Delano Roosevelt, spurred by Admiral Bullard, devised a plan for the Navy Bureau of Steam Engineering to guarantee Navy contracts to a new corporation that would endeavor to maintain an American presence in radio. This firm, organized by the General Electric Company (GE), bought out British holdings in the Marconi Wireless Telegraph Company of America, forming the new Radio Corporation of America. Initially RCA was merely a consortium to pool patents among the key U.S. players, but it also inherited Marconi's wireless telegraph transmitters and radiogram service in the United States (Aitken 1985).

Radio at the time used steam-driven, high-frequency electric alternators. GE was the major U.S. manufacturer of such equipment. The Navy Department was concerned about an offer British Marconi had made to GE to purchase exclusive rights to manufacture and distribute worldwide (including in the United States) the Alexanderson alternator—critical equipment in transatlantic radiotelegraphy in World War I. Defense interests—particularly Admiral Bullard—viewed control of these patents by a foreign power, albeit an ally, as a risk to U.S. national security.

The only generally accepted, commercial application of electromagnetic waves in 1919 was *radio-telegraphy*, primarily to enhance maritime safety. The sinking of the Titanic in 1912 had dramatized this application in the public mind. An international convention to regulate maritime radio traffic had been held in 1911 (a forerunner of the International Telecommunication Union's Comité Consultatif International des Radio Communications (CCIR), now known as ITU-R), with the Navy representing the United States. At the close of the war, the public was unaware of the secret wartime experiments performed by AT&T using vacuum tubes for transatlantic *radio-telephony*; indeed, conventional wisdom was that such techniques could not be perfected or commercialized.

So, RCA began as a wireless telegraph carrier and patent holding company. Alexanderson was its chief engineer while still an employee of GE. David Sarnoff, the U.S. manager of Marconi Telegraph, became RCA's general manager (Archer 1938; Dreher 1977). Soon after RCA's formation, other applications for radio technology besides maritime communications became evident. Unfortunately, three other firms held radio patents necessary for expansion of the technology: Westinghouse (which GE was then purchasing through a leveraged buyout), AT&T, and United Fruit. The latter had developed radio technology to improve command and control of its vast merchant marine and plantation operations in Central America. Each participant brought something essential to the table. Shares of RCA distributed among these three companies more or less according to the value of their technology. Initially the U.S. Navy held about 20 percent of RCA's stock, and AT&T was not added until some time later (Aitken 1985).

It is interesting to note that FDR's close relationship with GE's management, especially with its legal department, helped in the formation of the holding company.

But as often happens, little things create big changes. Despite the skeptics who were looking for applications in the wrong marketplace, commercial radio became practical and profitable within two years, but as radio *broadcasting*, instead of as a replacement for telephony (that came much later). This was a development RCA's original business plans had not anticipated. The first two broadcasting stations were AT&T's WEAF in New York City, and Westinghouse's KDKA in Pittsburgh (Banning 1945).

AT&T's radio engineers needed reliable transmission data to see if radio could be a useful mode of communication. To obtain that data, the engineers asked random listeners living in the New York metropolitan area to report back (by collective telephone of course) on the quality of reception of live and recorded music transmissions. One spin-off of AT&T's experimental efforts became Muzak. Another not insignificant outcome was the formation of WNBC radio in New York. The current National Broadcasting Company (NBC) network is a direct descendant of WEAF, which became WNBC. Most radios in those days were owned by dilettantes who listened to amateurs talking to each other, ships at sea, etc. Listeners benefited because they could enjoy music without having to turn on their gramophones. (Copyright restrictions did not even enter into the minds of the AT&T engineers.)

To make WEAF profitable, technicians working on the rooftop of AT&T's skyscraper headquarters at 195 Broadway started soliciting local restaurants and real estate brokers for paid ads. The first commercial was for an apartment house in Queens. The problem was that in the early 1920s AT&T's board did not see why funding radio research was worth the investment. They were also a bit scared that effective wireless would undermine their large investments in wires (Banning 1945).

Quite rapidly, AT&T built three separate networks of stations based on their Long Lines transmission system—the "Red" and "Blue" networks east of the Rockies, and the "Brown" network in the Pacific states. Other attempts by potential competitors were thwarted because they lacked access to quality intercity telecommunications lines. RCA, however, began a domestic radiotelegraph service that competed with Western Union and Postal Telegraph, and threatened to compete with AT&T's private telegraph lines as well. (AT&T went out of its way to assure stockholders that wireless communications was not really a threat to "wireline" circuits, which of course it was not in those days.) AT&T's Western Electric began to manufacture radio receivers in competition with GE. And television research began in earnest (and in secrecy) at GE, Westinghouse, and AT&T's Western Electric (Aitken 1985; Barnouw 1966; Barnouw 1968; Barnouw 1970; Danielian 1939).

By 1926, RCA had been transformed from a holding company (with a telegraph department) into a manufacturing subsidiary of GE/Westinghouse. The Navy and United Fruit had withdrawn, but AT&T's relationship was more complicated. From 1925 to 1926, AT&T and GE/RCA held a secret arbitration to settle a set of complex patent disputes. This resulted in AT&T's sale of its RCA stock, and an agreement that RCA would manufacture only home *receivers*, while AT&T's Western Electric would make only station *transmission* equipment. The Justice Department pressed GE to spin off RCA as a separate corporation. RCA, now under the direction of David Sarnoff, purchased the Victor Phonograph Corporation to gain access to a consumer distribution system. For a nominal fee of $1 AT&T sold its radio networks to the new National Broadcasting Company—a wholly owned subsidiary of RCA. NBC agreed to use AT&T's Western Electric manufactured transmission equipment and *only* AT&T Long Lines for its network connections (Danielian 1939).

More important, GE and RCA agreed not to enter the wireline telephone business, and Congress blessed this agreement by writing its principal terms into the 1927 Radio Act, and subsequently the 1934 Communications Act. Coming after some close calls with antitrust and nationalization in the prior decade, the 1927 blessing by Congress was important to AT&T. Its import had not gone unnoticed when FDR—

who after all had some acquaintance with both radio and telecom matters, at least from the GE/RCA side—became president in 1933. With the 1934 act, the government initiated a new inquiry into AT&T's anticompetitive actions (Cummings 1937).

Television research was transferred in 1929 from Westinghouse to a new RCA laboratory under Vladimir Zworykin. In 1926, AT&T also formed Bell Labs combining Western Electric's labs and the Long Lines radio research department. AT&T and RCA then competed for the sound movie business, but that is another story (Archer 1938).

While the original RCA did not perform quite as its sponsors had expected, the principal goal of this government/industry consortium—strengthening the incipient U.S. electronics industry—had been met. For the next half-century the United States dominated world industrial, military, and consumer electronics technology. And, during World War II, by order of magnitude, AT&T and RCA were the largest two military contractors on the Allied side (Fagen 1978).

AT&T sold its three nationwide radio/wireline networks (another brilliant innovation, since AT&T was a network company) for $1 to RCA in 1926 to avoid potential antitrust implications and a very messy patent conflict. RCA merged the Red and Blue networks into a new National Broadcasting Company. These legal/political actions moved the telephone company away from commercial electronic entertainment just as the new paradigms were being defined.

AT&T's model of a "central office" broadcasting music via localized radio waves, fed by long-distance (toll) wire circuits to supplement the public-switched telephone system, separated the "one-way" NII from "two-way" (interactive) NII for the greater part of this century, even though the initial reasoning of AT&T for a tariffed, common-carrier broadcaster, akin to the telephone, did not hold. "Convergence" took a back seat, but we are now back to square one in redefining the model for the evolving NII (Banning 1945).

RCA's role in establishing radio, and later in launching color television, cannot be matched by any one firm in the emerging convergent multimedia marketplace. Rather, economies of scale and scope must now be created through cooperation of diverse industries, firms, government agencies, and individuals. This daunting challenge can be met, if the principles of open communications policy are applied systematically.

We next contrast the traditional public trustee model of broadcasting and the common "calling" model of telephony and telegraphy with the more dynamic and flexible model of open communications that is now emerging and pushing convergence. All that came before was mere prelude to the activities launched by consolidating regulatory control in one agency that was to concentrate on communications issues for the next sixty years.

The 1934 Act and the Public Trustee Model

The 1934 Communications Act has remained for more than sixty years the principal legislative framework for communications in the United States. As we have noted, the evolution of communications legislation took place over a quarter-century, dating to the Mann-Elkins Act of 1910 which established the federal government's power to regulate telephones and telegraphs through the 1927 Radio Act which put some order into the chaos of anarchic spectrum interference.

The 1934 act was more of a consolidator and codifier of twenty five years of turmoil in communications than it was an innovator. As such, it established the Federal Communications Commission, an independent regulatory agency, to oversee the wireline and wireless communications industries. The FCC took over from the 1887 Interstate Commerce Commission, which had been obsessed with addressing the agonies of the railroads, which faced new competition from motor trucks and from overbuilding, and from the 1927 Federal Radio Commission, with its even narrower focus on spectrum allocation. Given the FCC's early legislative origins at the dawn of wireless telegraphy—a decade before radio broadcasting was introduced, thirty years before microwaves and television, forty years before the transistor and the computer, fifty years before microelectronics, sixty years before satellites and fiber optics, and seventy years before cellular radio and the Internet—we believe its time has come, and gone.

The commission, according to the 1934 act, was to ensure that the "public interest, convenience, and necessity" was served by private sector broadcasters and telecommunications service providers. These vague terms dating to the 1887 Interstate Commerce Act regulating railroads may still be relevant, but the regulatory practices are not only irrelevant to the challenges of the future, they are archaic and a waste of resources when dealing with logical networks based on digital architectures rather than physical infrastructures. If our focus is to shift to an open communications policy framework, we believe a more transparent policy process would limit legal delays stemming from endless regulatory proceedings.

The key principles and language of the act itself were taken almost word for word from the 1887 legislation establishing the principle of common carriage for railroads. The principle of localism—that broadcasters were to serve the particular interests of the community in which they served—was also explicitly identified in the act as a key measure the FCC was to use in assessing the performance of broadcasters. Other uses of the frequency spectrum that were not intended for use by the general public would not be held to the same standard, but would nonetheless have to justify themselves as being in the public interest.

While the Kingsbury Commitment established the public service precepts that broadcasting was to follow, it was for quite different reasons. In common, both the telecommunications service provider and the broadcaster were accorded gatekeeper, if not monopolistic, status in their respective markets. For AT&T, it was a monopoly blessed in one way or another by the federal and state governments in exchange for managed cross-subsidies between business and other customers in order to facilitate universal access to the public telephone network (at least in the more populous regions of the United States). For telephony, the legal underpinnings for this so-called "natural monopoly" model were based on ancient Anglo-American suppositions on the rights and duties of firms that follow "public callings," strictly regulated, in theory if not in fact, as "common carriers" (Cummings 1937; Horwitz 1989).

Broadcasting was different. While the Kingsbury social contract model held for radio service in exchange for state-sanctioned protection against *interference* on select spectrum (at least a natural monopoly for that part of the spectrum), the reasoning behind this model was based on analogies to riparian, or water rights, instead of those of carriers of commodities or purveyors of public utility infrastructure. Indeed, the plan proposed by AT&T for making broadcasting just another common carrier, with radio stations equivalent to telephone central offices, and radio time being "rented" under a tariffed and rate-regulated regime, as if it were merely a one-to-many telephone call (a concept that dated back to Alexander Graham Bell), was explicitly rejected by the federal authorities in the 1920s (Pool 1983).

The concept of a "public trustee" for scarce spectrum was invented for radio. As with common carriage, radio was still "in the public interest," but it was recognized that the airwaves dealt with content, and in the words of the Fourth National Radio Conference, which represented the nascent broadcast industry in 1925, that "the doctrine of free speech be held inviolate" in any legislation or regulation (Pool 1983). First Amendment issues were used (perhaps as an excuse with strong parallels to today's arguments by telecom providers seeking to enter or control content) to distinguish two-way from one-way communications. Pool writes "as the number of stations grew and interference worsened, the pressures for regulation grew. In the design of a regulatory scheme, among the most debated issues was censorship" (Lewis 1991; Pool 1983). Democrats were worried that a regulatory body dominated by a Republican president would restrict their access to the airwaves; the ongoing Scopes trial suggested to some that the government should prohibit "any radio broadcasts concerning the theory of evolution," but this was rejected by Congress. Content-dominated radio was clearly a different animal than telephones, where connectivity established value.

After a century of being more or less ignored, First Amendment ideals rose to new heights in the 1920s. Arguably, this may have been idealism: as a useful underpinning for democracy in reaction to the egregious abuses of World War I, as a bulwark against growing international fascism and bolshevism, or as a stabilizing factor during a time of economic distress, which began well before the 1929 crash.

The drafters of the Constitution put in the Bill of Rights the simple phrase that "Congress shall make no law ... abridging the freedom of speech, or of the press" for the simple, common-sense reason that they could think of no way of restricting speech without creating even more insidious difficulties with "the right of the people ... to petition the government for a redress of grievances." The late Professor Ithiel Pool at MIT used to tell his students that "no law means *no* law." Even the slightest deviation from "no law" tends to enter into the quagmire of attempted thought control. Unfortunately, it appears that every generation and every political regime has to learn this lesson the hard way by first attempting to restrict some speech that they do not approve of and then learning how extremely difficult it is to draw boundaries (Pool 1983, *passim.*). Senator Exon's obsession with pornography on the Internet and his insertion of an obviously unenforceable and unconstitutional censorship provision in the 1996 Telecommunications Act confirms the convictions of our founders, and the two centuries of successive court decisions, that an absolute prohibition against government interference in speech is the best practice in a true democracy (S. 652, 104th Congress).

Justices Holmes and Brandeis' famous dissent in *Abrams v. the United States,* a watershed case of seditious libel, was a sobering revelation of the power of speech in the hysteria of the postwar period (Lewis 1991). Holmes and Brandeis felt that the growing realization that the new radio technologies were potentially as dangerous as the printing press, more complex in function, and more pervasive, made the negative mandate of the First Amendment, "make *no* law...," a more attractive rule for this confusing new technology of radio broadcasting than that of the proscriptive language found in public utility legislation dealing with tariffs. Words are not like commodities and radio networks do not exhibit the characteristics of long- and short-haul transport links (Solomon 1989a).

The Radio Act of 1927 did contain language forbidding discrimination based on political speech—the "equal time" provision that is still on the books today—and prohibiting censorship except for profanity and obscenity. (Furthermore, it contained an obscure section, number 13, which restricted radio licenses from any person or firm found guilty of monopolizing radio communications, a tangential reference to the secret pact between AT&T and the Radio Corporation of America dividing two-way wireline from one-way wireless communications.)

The public interest was therefore perceived to be served differently for wireline (which meant telephones) and wireless (which meant broadcasting). So, in exchange for its exclusive license to use a portion of the spectrum for a particular purpose, the broadcaster was to act as a public *trustee* in its use of the scarce resource—frequencies—just as the owner of land that contains scarce water resources necessary for the community's survival is merely a trustee for the water, and cannot control its use only for his own whim or benefit.

The broadcaster could discharge its responsibilities as a public trustee by airing news and public affairs material, as well as programs for children and stories of local interest. The telephone service provider, on the other hand, could discharge its responsibilities to the public by acting to promote universal service, and by carrying all (telephonic) communications without discrimination as to rates or service.

A calm descended after the passage of the 1934 Communications Act. The act codified a quarter century of rules and sometimes imprecise understanding among the industry, customers and consumers, and the government. Under the act, the new FCC created a stable marketplace for radio broadcasting and telecommunications services that lasted until the introduction of computing technology and competitive long-distance carriers in the 1960s and 1970s. There were nonetheless fierce struggles to develop and control broadcast technologies, which presaged the fierce struggles we see today to develop and control the digital computer technologies that are replacing both older radio and telephone technologies.

The breakup of AT&T in 1984 resulted from digital transformation, the first strong winds of change blowing through the telecommunications industry since the early part of this century—changes that are now buffeting broadcasting as well as telecom. What we are witnessing is the creation of a new model, one that will most likely apply uniformly to both telecommunications and broadcasting. This model will replace both the public-trustee and public-interest models developed in the legal structures that were created with the legitimization of federal regulation over interstate commerce in the 1877 Munn decision and the 1887 Commerce Act, the 1890 Sherman and 1914 Clayton Acts to control the abuses of monopolies, the federalization of telecom regulation in the 1910 Mann-Elkins Act, the 1913 Kingsbury Commitment for universal service, the failure of Wilson administration to permanently nationalize AT&T in 1919, the 1921 Willis-Graham Act which partially reversed Kingsbury and encouraged further telephone consolidation, the 1927 Radio Act that brought order out of spectrum chaos, and finally the 1934 Communications Act that codified all that came before.

The turbulence of the past twenty years in telecom and broadcasting was presaged by recurring struggles for control of key broadcasting technologies in the 1920s and 1930s, such as the controversy over the patents for continuous wave broadcasting (Aitken 1985; Archer 1938; Solomon 1989b), frequency modulation (FM), color television (Crane 1979), and most recently, high definition television (McKnight, Neil, and Bailey 1995). The tragic fate of the inventor Edwin Armstrong—who committed suicide following a decade-long battle with the RCA cartel—serves as a metaphor for the stupidity of fighting over proprietary technology when greener pastures await expanding markets. Armstrong's widow finally won legal recognition and the bulk of RCA's patent revenue from television (due to a technicality in the use of FM sound in television broadcasting) for his seminal technical contributions to the development of amplitude modulation (AM) and FM broadcasting techniques.

We are suggesting that the new model for telecom must embody the Open Communications Infrastructure made possible by the use of a digital information architecture for all modes of communications, something that was not possible or even imagined some eighty years ago when the modern age of electrical communications began. OCI will relegate proprietary technical conflicts to the back room, opening up the vastly larger and more profitable markets of content and value.

Unless there is some recourse to OCI as the future model, business rivalry based on proprietary technological interests will repeat the struggles for control in the early twenty-first century—most likely at a more intense level than ever seen before; digital technologies are like that: either completely open like the Internet, or disastrously closed so that nothing eventually works. As in all battles, both Murphy's law and the law of unexpected consequences come into play, and the outcome is less certain than if there is agreement in advance as to the rules of the game and the logical parameters for competition. This will be particularly true as the value of content becomes paramount and the value of carriage lessens in all-digital information architectures.

In the 1990s, service providers work with a variety of technologies, and broadcasters reach two-thirds of their audience through multichannel cable television systems. With the advent of competitive local telecommunications the rationale for the public-trustee model as the most equitable to allocate scarce spectrum is fatally weakened in this reconfigured marketplace. However, a new model that adequately balances the First Amendment issues of access to information in an era of digital information architectures (a now greatly expanded amendment some seventy years since the Brandeis and Holmes dissent) with antitrust considerations (also greatly expanded since the Sherman and Clayton Acts), and more mundane issues of access to capital, problems with balance of trade and foreign domination of critical information technologies has not been devised to replace the public-trustee concept.

The use of spectrum auctions is the final measure to challenge the public trustee model. We will discuss this following a brief review of broadcasting from the 1930s to the 1990s.

The Color Television War

Whereas the battle to control radio broadcast technologies took place on an international battleground, the development of black-and-white television and later color television was initially a domestic matter. The battle fought domestically and internationally was for exclusive rights to technologies and frequencies. The framework for an open communications policy was far from the focal point of the wars.

The evolution of technical communications standards often makes for interesting history and provocative policy analysis (Neuman 1991). Rhonda Crane (1979) demonstrated this in her fascinating study of the European "color TV wars" of the 1960s and her development of a national champion model of standards debates. The United States expected the rest of the world to adopt its technology (NTSC-Color); the French developed a slightly higher resolution system with a radically different color technology (SECAM); the Germans, objecting to French royalty demands, developed their own color system (PAL) similar to the American NTSC model, but with the higher resolution of the French system. The British had initially indicated they would adopt a high-resolution version of NTSC-Color, but switched to the German PAL system to keep the French from dominating with SECAM.

President Charles de Gaulle and the full force of the French state lobbied the international community in favor of SECAM. In 1965, after providing the Soviet Union with royalty-free access to SECAM technology for local manufacture, and after enlisting the support of former French colonies, France was able to gain a bare plurality of the votes at a meeting of the Consultative Committee for International Radio (CCIR) of the International Telecommunication Union needed to adopt SECAM as the designated world standard—a ruling that has no force anywhere in the world where it has been ignored, which is to say, most of the world (Crane 1979).

It is hard to say which of the three systems is "better" or "worse," because each one has superior and inferior features both in picture quality and technical or economic externalities. Nevertheless, the systems are different, and eventually the world ended up with thirty, somewhat incompatible, variations of the three basic methods for transmitting color TV over-the-air, not to mention another set of standards for recording on tape and for interfacing with cable and satellite transmission. Neither the United States, with its sunk investment in NTSC, nor West Germany, with its domestic electronics industry export interests, were willing to abandon functional color television technology for the French variation.

The multiplicity of standards surely has raised costs for the consumer for international program distribution besides reducing picture quality for conversion. It has also served as a nontariff trade barrier, discouraging the international exchange of broadcast programs. Moreover, the Soviet bloc countries clearly wanted access to the French technology for a variety of reasons, and they were not at all concerned that initially SECAM sets in Eastern Europe could not receive PAL broadcasts from transmitters just across the Iron Curtain (and to make sure there were no leaks, the Soviets reversed the polarity of the color signal, so even distant signals from France would not be easy to view on the early TV sets; this was before VCRs and solid-state electronics changed the paradigm).

Crane's analysis showed, however, that French industry profited immensely from having a proprietary interest in the SECAM sets and studio equipment sold in the few areas that accepted their system, as well as by not having to pay German or American patent royalties for sales in France.

If the world does not adopt open interfaces for future video standards, we see history repeating itself for networked multimedia, despite all-digital infrastructure and connections. Already we have dozens of incompatible set top boxes being proposed for cable and satellite distribution, and multiple proprietary video compression standards that will raise the cost of maintaining electronic libraries of programming and films, but that also have the potential for concentrating at least short-term profits for those who can capture significant markets early in the game. Costs for the broader socioeconomic sphere are of course ignored in such shortsighted thinking. But there have been some exceptions. One has been the transition from very high frequency (VHF) to ultrahigh frequency (UHF) television broadcasting.

An OCI Precursor: The All-Channel Receiver Act

The 1958 All-Channel Receiver Act is not widely recognized to be the landmark that it really is: the first legislation mandating open interconnection of advanced information services. The services in this case were the recently licensed UHF broadcasters. They faced the classic chicken-and-egg dilemma: very few consumers had television sets capable of receiving UHF channels, so potential revenue from advertising was limited. Until most of the projected audience could receive UHF, their reach could not be extended, and therefore the attraction of their programs would remain low because programming finances were tight.

The All-Channel Receiver Act overcame this dilemma by mandating that all new television receivers sold in the United States be capable of receiving UHF broadcasts. The Commerce clause of the Constitution gave Congress authority to do this, given that all uses of the television spectrum clearly affect interstate commerce. This would have been much more difficult for telephony, where a convenient fiction that separates local and interstate characteristics has been accepted from the beginnings of its regulation.[47]

The All-Channel Act imposed an extra cost on manufacturers and consumers, because the electronics were more complex for the sets capable of receiving both frequency bands. The cost differential was on the order of $50 in 1960 prices. This was a price deemed by Congress to be worth paying to provide the long-term benefits to consumers of wider program choice—the First Amendment issue again whereby negative economic externalities are outweighed by the value of access to more speech. This logic is the reverse of the SECAM decision, whereby less speech was traded for economic and military advantage, in France and the Soviet bloc countries.

Similarly, today a number of questions are being raised by the radical shift to digital technology and digital information architectures. They range from the choice of future advanced television standards to be fully compatible with multimedia and computer displays, to interfaces with varying local access infrastructure. One example we have struggled with is: Are the costs involved in requiring progressive scan transmission techniques by future high-definition TV broadcasters worth paying earlier for long-term economic and social benefits expected from accelerating the convergence networked interactive multimedia services and applications?[48] Although some manufacturers argue that the costs imposed are excessive, the value to society of a strong information sector and expanded communications capabilities are potentially far greater, as we discuss below.

Before we address the future of broadcasting, we must understand the development of the multichannel marketplace, that is, cable television—one alternative from which a National Information Infrastructure may emerge.

From Community Antennas to Cable

Cable television, like many other new technologies, was not recognized for many years to be what it is, a source of multichannel television. Rather, the "community antenna" that would receive programs and bundle them on to a wire was the significant point in the minds of its earliest adopters—small-town entrepreneurs in valleys, who could not receive distant television broadcasts due to topological and geographic factors. By pooling resources, an extra-large antenna could be placed high to receive the distant or weak signals, and feed the signals into a cable for home delivery. In spite of its humble origins, in the 1960s broadcasters came to fear the potential competitive threat of cable television. Public campaigns to "save free TV" were mounted, and succeeded for a time in rallying public support for broadcasting-industry lobbying to stunt the growth of cable television. Federal Communications Commission regulations required that only old movies could be shown on cable channels.

Because the FCC considered community antenna television systems to be merely a vehicle for broadcasters to improve the quality of signal reception, cable firms were considered to have the royalty-free right to redistribute signals received terrestrially through their systems. Whereas cable system operators were small and weak into the 1980s, the three broadcast networks ABC, CBS, and NBC dominated. Therefore, to limit their control over the video marketplace, FCC rules restricted the number of programs they could produce for themselves. Most significantly, the FCC regulators prohibited the networks from sharing in the revenue generated from program syndication (the broadcasting of reruns); these rights were retained by the producer. As the networks' share of the viewing audience declined from over 90 percent in the 1960s and 1970s to just over 60 percent in the 1990s, the urgency with which broadcasting industry lobbyists sought to have these rules—known as the financial and syndication rules, or fin-syn—increased. However, despite the powerful economic interests of Hollywood studios and producers, which had prospered with the protection of the fin-syn rules, court proceedings and rule makings from 1992 to 1995 led to the rules being formally repealed.

With the commercial development of satellite distribution to cable headends, an economic means of national broadcasting through cable television systems emerged. As the cable distribution system grew in importance, rather than limit cable systems, broadcasters wanted assurances that their signals would not be excluded from the systems by the cable operators. The broadcasters wanted a legal right to have their signals carried on local cable systems—that the systems "must carry" them. The "must carry" regulations have come and gone and come again. The issue of the interplay of the market power of the cable operators and the First Amendment rights of broadcasters and cable operators is complex. A surprise Supreme Court decision in 1997 threw this issue back into play. Suffice to say that the economic threat to local broadcasters is very real, when 60 percent of the viewers in their region receive television through cable systems. Exclusion from the cable system in effect reduces the potential market reach of a broadcaster by two-thirds.

As noted, cable systems enjoyed royalty-free rights to programming. In addition, with the elimination of the must-carry rules, they gained the power to exclude broadcasters, which they would typically do only to the lowest-watched stations. With cable regulation, the tables have turned yet again, and the cable operators are now obligated to compensate broadcasters for the use of their programs.

Cable operators in general refused to provide cash compensation, and instead have bartered channel capacity for programming. Instead of paying a broadcast network, the network's parent company is offered an additional channel to use for other programs.

The Paradox of HDTV

The paradox of high-definition television stems from the fact that the struggle to control its development has very little to do with television as traditionally defined. Rather, nationalist interests, corporate struggles to maintain global economic competitiveness, and the paradigm shift to a model in which every viewer may also be a producer have taken the battle over HDTV far beyond the production of prettier pictures for home television viewing. Furthermore, with a shift to digital television technology, which became a pitched political battle in the 1980s, HDTV modulation suddenly became useful for all types of digital transmission, and the broadcasters are now asking to be allowed to use their future all digital frequencies for services potentially more lucrative than TV, much less high-definition TV.

HDTV was to be a better kind of television, a means for displaying entertainment in homes and perhaps also in movie theaters. The effort to win international acceptance of an HDTV standard was led by the Japanese public television broadcaster NHK in concert with Sony, CBS, and the U.S. Department of State. This international public-private partnership began the technical debate on a new HDTV standard in the 1970s and unraveled following its failure at the International Telecommunication Union's Consultative Committee for International Radio (CCIR) plenary meeting in Dubrovnik in May 1986.

In late 1982, the Advanced Television Systems Committee (ATSC) was formed. The stated purpose of the ATSC is to recommend positions for the United States' use in international standards organizations. Ironically, ATSC rapidly became dominated by non-U.S. manufacturers.

Professor William Schreiber of MIT strongly criticized the NHK system early on, arguing that their system did not take advantage of developments in digital technology, semi-conductors and signal processing, and that it required too much bandwidth: "[a]s a result its system design is no more advanced than that of the 32-year old NTSC [color] system" (Schreiber 1988).

The NHK system was characterized by analog processing and an interlaced picture field, almost identical in concept to the low-resolution systems still used worldwide for television. Interlace produces annoying artifacts, and the low field rate proposed produced flicker that can sometimes result in some individuals getting headaches and feeling queasy. As a result of these problems, the computer industry dropped the low field rate and interlace very early in its adoption of display devices. The main criticism of the NHK system therefore was that it was not compatible with whatever was evolving on the computer front, and it was unnecessary to propose something that might be cheaper in the short run if it was going to raise costs and compatibility problems in the long run. But underlying these criticisms was another problem: some American interests and the European Commission opposed adopting a future technology that depended solely on patents and technology controlled by Japanese firms.

In spite of the criticism, the State Department's Bureau of Communication and Information Policy and the ATSC managed the policy process throughout this period. The goal was for the 1986 CCIR plenary conference at Dubrovnik to accept the 1125/60 standard as a single worldwide standard for production and international program exchange. The ATSC was able to present the State Department with a unanimous recommendation for the new system (Mestmaecker 1987), despite behind-the-scenes grumbling from disappointed advocates of progressive scan.

The State Department/ATSC position was that agreement on a universal standard would benefit U.S. international trade and program producers in particular, because the United States had the leading program production industry in the world. Export of U.S. produced programs is an important revenue source (see table 5.1). It also happens to be one of the country's most successful exports. The ATSC and the State Department, however, had not thought through the difficulty of implementing a new incompatible standard. Most producers were in no hurry to switch to a new format that in the first instance would reduce the size of the

Table 5.1
U.S.-produced programs broadcast per day in Europe

Country	Percent total broadcast time
France	16
Germany	25
Italy	31
United Kingdom	24
European average	25

Source: Suzanne Neil, based on U.S. Department of State/ATSC, "A Worldwide High-Definition Television Studio/Production Standard," unpublished report, July 1985.

market they could serve to near zero—that is, to the very few HDTV outlets available, which were largely trade shows and test transmissions.

The Dubrovnik Debacle
Before the wars that shattered the Balkan peace in the 1990s began, the world's television engineering community had already split into factions while meeting in Dubrovnik in 1986. The major U.S. 1125/60 supporters worked closely with Ambassador Diana Dougan, director of the Bureau of Communication and Information Policy at the State Department, and vigorously lobbied both the Europeans and the Australians to accept the 1125/60 proposal. The lobbying efforts, however, were unsuccessful: the Europeans, led by officers of the European Community Directorate General 13 on Innovation in Information and Communication Technology in concert with representatives of Thomson, Philips and Siemens, united in opposition to the U.S.-Japanese position. The result was that the CCIR plenary agreed to defer choice of an HDTV standard until the following CCIR study cycle (1986–90).

The defeat in Dubrovnik left the State Department and ATSC officials—and the 1125/60 Japanese proponents—in a quandary. Ironically, vigorous U.S. lobbying efforts in Europe and Australia had the effect of catalyzing European opposition to *any* U.S. or Japanese standard, and therefore to a single global standard. The assumption was that if Japan and the United States wanted the standard adopted so badly it must be bad for European economic interests. It also inhibited—though did not halt—U.S. research efforts or stop entrepreneurs from proposing alternative technologies and standards.

The HDTV debacle in Dubrovnik illustrates the continuing short-sightedness of the players from both the public and private sector who fixated on winning a narrowly defined standards battle while ignoring the broader technical evolution of digital communication. Ironically, U.S. research and development of advanced television was discouraged by the State Department which insisted that the only issue was how to get the new standard adopted. To the State Department, the debate over interlaced versus progressive scan was a minor technical quibble. It was, in fact, fundamental to the future of digital information architectures.

The FCC Takes the Lead

The partnership between the Federal Communications Commission and the broadcast television industry has roots that go back to the invention of television. So it was no surprise when the partners determined the definition of HDTV and who would be involved in developing it. Because that definition was narrow, the selection process excluded related technological developments (mostly in computers and telecommunications) and industry representatives. In the end, the HDTV technologies under FCC consideration had changed (Setzer and Levy 1991).[49]

The FCC became the central government agency in the HDTV debate through its authority to regulate the radio frequency spectrum. The State Department's loss of face following the debacle in Dubrovnik silenced its Bureau of Communication and Information Policy on HDTV matters, although its role as head of U.S. delegations to international standards meetings left it with some residual influence. The immediate cause of the FCC's interest was the petition of land mobile communications companies for spectrum allocation at the 12 gigahertz band (General Docket 85-172). As a blocking motion, the Association of Maximum Service Telecasters (the terrestrial television broadcasters' association) and fifty-seven broadcast companies filed a "Petition for Notice of Inquiry" on 13 February 1987. The broadcasters wanted the FCC to open the question of the impact of new broadcast technologies (both television and mobile radio) on FCC spectrum-allocation and channel-assignment policies.

The FCC's tight link with broadcasters was based on its regulation of the broadcast industry. In the name of protecting the public interest, the FCC also protected the broadcasters. The definition of HDTV was narrow and ... unquestioned: it meant an updated form of television for home entertainment. The "world view" of this debate was also narrow; the discussion was about the U.S. market; the tacit assumption was that equipment came from other countries, particularly the Pacific rim, and that the global dimension existed only in terms of international program exchange.

ACATS: The Advisory Committee

To help advise the FCC, the notice of inquiry (NOI) called for the creation of a citizens' advisory committee, known as the Advisory Committee on Advanced Television Services (ACATS), which was impaneled on 17 November 1987.[50]

The commission, in its "Tentative Report and Further NOI" issued on 1 September 1988, restricted broadcasters to 6 megahertz channels in the currently assigned UHF and VHF bands; required that current National Television Standards Committee (NTSC) users be able to receive advanced television (ATV) material (although it found simulcasting acceptable); and postponed FCC consideration of intermediate standards until after an ATV standard had been chosen. This critical decision was unusually farsighted and had a tremendous impact on the future development of advanced digital imaging technologies and the emergence of the broadband media system.

The major issues HDTV raised before the FCC were how to allocate spectrum among competing users and whether and how to support terrestrial broadcasting in an evolving marketplace for advanced television services—what is currently referred to as multimedia. Debates over the possible allocation of new spectrum for HDTV, and over the size of channels allocated to the new television service, were heated and confused given the differing conceptions of how television would be used in the future, including debates on cable and satellite HDTV policies, receiver compatibility, and cost of HDTV retrofitting. It turns out that, as it was in the beginning, this debate is once again the core regulatory issue surrounding HDTV: that is, whether to permit the TV spectrum to be used for something other than conventional, over-the-air television.

"Free" terrestrial television had been facing growing competition over the previous twenty years. The competition came from the increasing strength and number of other electronic networks, primarily cable and satellite broadcasting and video entertainment systems, especially VCRs, but also video disks. In 1970, as work was just beginning on HDTV, terrestrial broadcasters dominated well over 90 percent of the U.S. television market. By September 1988, when the FCC released its tentative report, terrestrial broadcasters had the minority share in their competition with cable, although cable routinely retransmits the local over-the-air signals, and VCRs could be found in over 60 percent of American homes.

Alternative Systems Proposals
In 1987, William Schreiber and Andrew Lippman at MIT began a new debate on receiver design for advanced television with a proposal for an "open architecture receiver," a device that embodies the essential OCI principles. They argued that given the rapid advances in all aspects of video technologies, it would be most useful to build flexibility into receivers. They suggested building a receiver using the bus architecture common to personal computers, with open interfaces between the core of the receiver and the "peripherals." Furthermore, Schreiber called for the creation of a "friendly family" of standards, a series of interrelated standards, each of which had different levels of quality or were optimized for a different transmission medium. Such a development would give receiver manufacturers economies of scope and scale and would provide the end user (home viewers) with easy access to technological improvements as they became available.

The Role of Congress
The public debate on future broadcasting systems also began in earnest in September 1988, at the time of the discussion of the FCC's tentative report and further notice of inquiry, with remarks by Edward Markey, then chairperson of the House Telecom Subcommittee, at the opening of hearings on HDTV. Markey argued that the future of HDTV in the

United States would influence the state of U.S. consumer electronics as well as U.S. advances in microprocessing, photonics, and other related industries; it would affect tens if not hundreds of thousands of jobs; and over the ensuing decades, it would involve $50 billion to $250 billion worth of economic activity. Markey stated that advanced technologies are the future of the electronics industry. He also noted that without reentry into the area of consumer electronics, it will become increasingly difficult to defend America's main economic strength—technology.

As chair of the Telecommunications Subcommittee, Markey was a natural focal point in Congress for HDTV activities, although he was not the only one. George Brown, chair of the Science, Technology and Competitiveness Subcommittee of the House Committee, and later as chair of the full committee, also played a central role. In the 1988 Omnibus Trade Act, Brown had succeeded in including a provision for creating an Advanced Technology Program (ATP) in the old National Bureau of Standards (now NIST). This was the first in a series of legislative actions aimed at supporting HDTV as a means to increase U.S. international competitiveness.

The Role of ARPA and the Aborted Role of Commerce

Both the executive branch and the computer industry also became publicly involved with HDTV in the fall and winter of 1988–89. The first executive branch action was the release of the Department of Defense's Advanced Research Projects Agency (DARPA) Broad Area Announcement, calling for proposals for the development of digital signal processing and display technologies, and awards totaling $30 million in grants for technologies that would have civilian as well as military applications. Eighty seven submissions came from large and small firms covering display processors, receivers, display technology, as well as other technologies. As HDTV develops, these elements will be critical to its success, something that the early proponents, including the State Department with their lack of technological prowess, tended to ignore.

The Commerce Department also became involved in HDTV. Out of the public eye was the question of funding for the new advanced technology program. One of Robert Mosbacher's first statements as George

Bush's new secretary of commerce focused on HDTV as a central trade and industrial development issue. "HDTV is not just another stage in TV ... not just another consumer good, but a whole other generation of electronics" (*Communications Daily* 25 January 1989). It was unclear what specific form Commerce support would take, but the options included government funding for a research consortium such as Sematech, as well as expanding NIST involvement in the measurement, testing, and certification of HDTV standards. This represented an extension into a new domain of NBS's traditional involvement in standards setting. One of Mosbacher's top aides, Assistant Secretary for Technology Bruce Merrifield, admitted that the government would have to provide "seed money," because it was highly unlikely that any cooperative venture would be entirely privately funded (*Communications Daily*, 30 January 1989).

In May 1989, the American Electronics Association (AEA) released a proposal for a new public-private initiative, using HDTV to leverage U.S. recovery into consumer electronics. The plan covered a ten-year period and called for easing antitrust restrictions, formation of a public-private sector partnership, and federal subsided loans (which would be repaid), totaling $1.36 billion over the life of the program. Suddenly, there was a shift from the administration: Mosbacher publicly characterized the $1.36 billion subsidy as a handout from "Uncle Sugar," thereby coalescing all the opposition to federal activity around one politically appealing if inaccurate point (the figure was a loan with a repayment plan in the AEA proposal). Lobbyists for the NHK system and the press picked up on the handout issue and never let go.

The spring of 1990 marked a turning point in the development of both a U.S. and an international HDTV standard. The State Department, as noted above, had finally agreed to withdraw U.S. support in the CCIR for the Japanese proposal; the Department of Commerce Advanced Technology Program was funded and had begun its first grant process; and the Committee for Open High Resolution Systems (COHRS) came into existence. The HDTV debate widened to include the computer industry, as well as others with an interest in the standardization process. Internationally, the CCIR at the May 1990 Dusseldorf Plenary postponed for yet another four years a decision on the standard. The creation of both COHRS and a CCIR Task Group on Harmonization, and the study of digital HDTV and alternative systems within ACATS, were all motivated by the growing potential of digital technologies to produce, transmit, and receive images and sounds.

The DARPA HDTV grants were oriented toward digital signal processing and digital displays. William Schreiber and Andrew Lippman's proposal for an open architecture receiver, built on the technology of personal computers, made the link between HDTV and computers explicit and pointed to a new area for development. Computer companies began testifying before congressional committees and responded to FCC notices of inquiry concerning HDTV. Under the auspices of the U.S. chapter of the Institute of Electrical and Electronic Engineers (IEEE-USA) and the Advanced Television Systems Committee, academics instituted a series of off-the-record meetings between computer industry leaders and television interests in order to build bridges between the two worlds.

The Committee for Open High Resolution Systems
The establishment of the Committee for Open High Resolution Systems in the spring of 1990 marks the beginning of a new attempt to define an HDTV standard. As its name suggests, this organization, which never had a formal structure or membership, was dedicated to finding a way to build open video systems. The members, who were largely from computer companies or academic institutions, were agreed that it would be desirable to create advanced video systems, including HDTV and digital television, that (1) would be useful in different environments (computer graphics, television, medical imaging and healthcare, military training, and computer-aided engineering, to name a few); (2) would be amenable to upgrading; and (3) would be useful at different levels of refinement. The terms for these qualities are "interoperable" (across industries and applications), "scalable," and "extensible" (across time and degrees of refinement). Interoperability, for example, would be of interest to manufacturers who could capture economies of scope in manufacturing a range of related products, from television sets for home entertainment to monitors for air traffic control. Scalable systems permit the receiver to select only information as needed from the signal; therefore, the same signal could be transmitted to receivers with different levels of sophisti-

cation. Extensible systems are ones that can incorporate technological improvements; extensibility refers to the ability of the system to incorporate new technologies over time.

Also supporting harmonization efforts were a series of meetings, beginning in June 1990, that were sponsored by the Institute of Electrical and Electronic Engineers and co-sponsored by the ATSC, on digital video systems. At the first meeting, after some debate and heated accusations from frustrated NHK supporters, the participants agreed to create several subgroups that would be responsible for ongoing work. One was the "header group." Its task was to design a digital identifier or "header" that would contain all the information necessary to permit scalability and extensibility in video signals.

The feasibility of developing a standards architecture that would be modular, scalable, and capable of extension with new technologies— some not yet imaginable—was considered in the late 1980s and early 1990s by task forces encouraged by government officials. These included computer and telecommunications industry and academic experts. Representatives of a few broadcasters and cable firms also participated. The objective was to develop an interoperable digital image architecture to permit the seamless movement of bits across motion picture, broadcasting, cable, computer, satellite, wireless and wireline telecommunications networks, and medical imaging industries, to name just a few sectors that at that time were still quite disparate. It was with a giant leap of faith that this effort began when the World Wide Web was still just a gleam in software engineers' eyes, and video, much less high-resolution video, on packet networks like the Internet, was disparaged by television executives. Progress in defining an extensible, scalable, and interoperable architecture was made by International Telecommunication Union, International Standards Organization, Internet Engineering Task Force, and Institute of Electrical and Electronics Engineers committees, and especially within the broadcasters' own standards organization, the Society of Motion Picture and Television Engineers (see SMPTE, 1992). The key to this success was found in the principles of ad hoc virtual governance and agile policy response, as we described in McKnight (1989) and McKnight and Neuman (1995).

A major breakthrough was the adoption by SMPTE of the flexible "header-descriptor" format later mandated by the FCC for digital television broadcasting. This digital architecture permits all future bitstreams to be readily identified, so that an imaging stream could be decoded, manipulated, and viewed on future display devices. A standard could be decoupled to allow more graceful technological evolution. It is a truly revolutionary architecture, one designed to move us beyond the debilitating interfirm and intersector battles over standards.

The impact of these task forces was to shift the FCC process on advanced television from proponent systems that were only analog or hybrid analog/digital to all-digital proposals. Following the lead of the computer industry as expressed in the COHRS efforts, on 15 June 1990 General Instruments (GI) submitted the first proposal to the FCC based on digital signal capture, transmission, and display. Within six months, all of the remaining proponents except NHK introduced proposals for digital systems. By May 1993, all of the digital proposals, with the encouragement of the FCC, were merged into a "Grand Alliance," save for the Japanese who eventually abandoned their analog HDTV system.

By the mid-1990s, HDTV was finally on a path of cooperation with other imaging industries. But it was not to be. As we write in the spring of 1997, there is yet another initiative to broker a progressively scanned video signal standard common to the computer and television worlds. If the Grand Alliance were truly a grand alliance, why would this be necessary? A brief review may make this clearer.

The Japanese and Europeans, through the 1980s and early 1990s in their haste to rush their HDTV systems to market, failed to heed the advice of advanced scientific research that could have been made available to them. In the United States, research was conducted at MIT, Columbia University, and a variety of large and small industrial laboratories. Audience research, for example, revealed that from a normal viewing distance of 9 feet from a television set with a 28-inch display, most untrained observers could hardly detect a difference between HDTV and the existing NTSC system. This was research the Japanese or Europeans could have conducted but did not.

Through this period, the Bush White House actively opposed government involvement in the HDTV debates for fear of picking winners and preempting the marketplace. Lobbyists working for European and Japanese firms and governments encouraged the dominant ideology of the free-marketeers. But the ad hoc, agile technology policy process and the informal networks that had been created among academics, industrial researchers, and government officials remained active and influential throughout the Bush administration.

Even when some of its most visible proponents lost influence, the ad hoc technology policy process continued unabated. FCC Chairman Alfred Sikes artfully dodged administration "no industrial policy" bullets, which felled ARPA director Craig Fields. Fields earned the enmity of the White House and foreign economic interests by publicly calling for a major U.S. technology development effort for what in colloquial terms today is called "the information highway." Fields' perceived insubordination led to his being encouraged to resign. Sikes, however, provided leadership to steer the FCC Advanced Television standards-setting process toward digital and interoperable HDTV systems and away from obsolete technologies in which the Europeans and the Japanese continued to invest. Key individuals at U.S. firms and research institutions with high stakes in digital imaging communications played prominent roles in this endeavor, working closely with Sikes and the FCC process, and with other government agencies. This is typical of the kind of ad hoc, agile technology policy the American political process helped create.

Despite administration opposition to industrial policy, ARPA began supporting flat-panel display consortia in the late 1980s. ARPA also supported development of magnetic recording devices needed for HDTV and mass computer storage. The distribution system in the United States for professional and consumer video equipment is dominated by foreign manufacturers, making it difficult for U.S. manufacturers with advanced and superior equipment to compete, even where the military would prefer a variety of sources for such devices. This is where government can help most by ensuring a level playing field that gives American firms a chance to enter the marketplace.

The evolution of digital television is not yet evolved, and the Grand Alliance of HDTV proponents is neither grand nor an alliance. HDTV was never about pretty pictures; it was about jockeying for market share in the evolving digital marketplace. Making free spectrum available to existing broadcasters to protect "free television" and make possible crystal-clear pictures and CD-quality sound reflects more the political power of the broadcasters than the technical power of the Grand Alliance compromise. The compromise is not a standard: It is a compromise, a list of eighteen different, non-interoperable standards, ranging from the original Japanese analog interlaced picture of the late 1970s (albeit reincarnated as a "digital" system) to one that the computer industry recognizes as fully interoperable with digital computer graphics. A particular irony is that the compromise package permits the allotted bandwidth to be used for multichannel conventional television, using the advanced electronics to squeeze five NTSC channels into the space previously occupied by one. This is an interesting development, especially if it results in advanced television as over-the-air multichannel cable.

The advanced digital television compromise offers no guarantee that the high resolution component on which the policy is justified will become a meaningful part of the broadcast picture. Inadvertently, however, Americans may be required to replace 270 million analog TV sets by 2006 at prices ranging from about $200 for a digital converter box, to $5,000 or more for a new digital large screen and standard resolution display. All this is a result of the attraction of free spectrum, not advanced television technology. Recall that for broadcast stations and networks, these new studios and transmitters add cost without necessarily any increased income from advertising.

The paradox continues and the knot remains firmly tied—the federal agencies, computer manufacturers, television equipment makers, the wireless and cellular telecom firms, and most prominently the broadcasters have struggled through two decades of HDTV wars. The public is puzzled and, as yet, not well served by this process. Perhaps little has changed since the robber barons fought with each other, the public, and the government over railroad routes and rates 100 years ago, only to see the industry superseded when new technology arrived in the form of the motor vehicle.

Computer Combat

Earlier we discussed the handicap that the Justice Department's consent decree placed on AT&T and telecom/computer developments in general.

It is worth delving into the background of computing a bit to understand how such blinders came to be so early in the game, and how a narrow view became received wisdom only to painfully fall by the wayside a few decades later.

International Business Machines reluctantly entered the computer era in the mid-1950s. Up until then, it had built an incredibly successful worldwide enterprise based on significant market capture of the tabulating machine business and associated appliances. Tabulating devices used paper cards with holes punched in them that represented numbers—and later letters—using a coding system similar to that for holes punched in paper tape for the Baudot teletypewriter and "ticker tapes" of stock market fame. This system of holes in paper, initially quite simple in conception, evolved to become a tool preventing interconnection between competitors' peripherals, thereby locking customers into one vendor, at first for tab-card devices and later computers.

IBM's direct corporate predecessor was Hollerith's Tabulating Company, originally established to build devices to sort the 1890 census—the first anywhere to be mechanized. The 1890 census was published only two years after it was taken, in comparison to the 1880 census, which was still publishing figures as the 1890 census was being organized.

By 1914, Hollerith's company was taken over by a young, energetic supersalesman, Thomas Watson, Sr., who promptly saw the global impact of automatic tabulating. IBM manufactured a range of office appliances, but their mainstay was Hollerith's tabulators. Tab machines tracked the finances and inventories for the burgeoning industrial complexes of railroads, petroleum distributors, and other basic infrastructure (AT&T's Bell System was its biggest private customer). Tab sorters were key for the operations of the New Deal's Social Security system and the growth of internal revenue services all over the world. By World War II tabulating calculators were harnessed to break codes and attack mountains of data for the Manhattan Project's atomic bomb designs—processes that required sorting and printing, with a little bit of adding. Toward the end of the tabulating era, some machines were able to follow complicated calculating algorithms, but tab-card calculators were not computers in the modern sense.

All during this period, IBM increased its market share because of three basic principles that later guided its entry and helped to maintain its hegemony in computers for decades:

• providing the customer total and complete service; a strong argument for
· forbidding non-IBM devices from connecting to their machines; and, because technology is critical in this line of reasoning
· maintaining control of the critical technologies.

These were enforced through leasing instead of selling devices, using trade secrets for details of interoperability, and engaging in constant research and development to assure a technological edge. These principles were very similar to those used by AT&T to maintain their legal monopoly over telecom; the main difference being that IBM *had* competitors and AT&T did not. IBM, however, did have some technologies that their competitors could not easily compete with, and they would not license them—a position that landed them into hot water with the Justice Department later on. One subtle, important, and closely held trade secret that guided the destiny of computing during its formative years was IBM's ability to manufacture tab-card stock cheaper and more accurately than any other competitor. So when they got into computing and again became the leader in their market, there was little incentive to skip the tab-card, batch-processing environment and go directly to more interactive systems.

IBM had built some experimental computers after World War II, using relays and vacuum tubes, but for the first decade of the stored-program experiments they did not take electronic digital computers seriously as a potential business. The stand-alone, special-purpose tab-card business was safe, very profitable, and growing. It was safe because of leasing and the absolute prohibition on interconnection—even for the supersecret crypto agencies during World War II, which had to hire IBM employees with proper clearances to maintain and work their machines. To protect the firm, IBM had a higher standard of security than the spies!

IBM's refusal to license their processes assured profitability. Business was growing because automatic tabulating was a route to productivity in a large number of sectors, especially in defense and manufacturing.

But tab-card machines were task-oriented devices, inflexible, difficult to modify, and had only very rudimentary programming capabilities.

Two separate but related happenstances in the early 1950s changed IBM's comfortable position as a manufacturer of precision tab devices, and brought the new world of automatic computing out of the engineering labs and into the world of business. In both cases, IBM had learned the lesson of their success well: whatever they did—punching holes or making sophisticated calculators—they knew they must keep total control over their market by controlling access to their devices and by segmenting the market so that they kept the greater market share in each segment as technology advanced. In the days of hardware dominance, this was a perfect strategy if they could keep the Justice Department at bay. But over the decades, software became their Achilles' Heel, something that not only did not matter much with tab machines, but that inventors did not even think was worth legal protection in the early years of computers.

The first seminal event was a patriotic effort by Thomas Watson, Jr., the son of the president of IBM, to help out the nuclear research establishment and the code breakers at a critical time in the cold war by building the first mass-produced stored-program machines. Watson's advisors, including one of the inventors of the digital computer, John von Neumann, estimated that less than a dozen machines would be sold worldwide, based on a "market" to solve all known mathematical problems. After that, von Neumann suggested everything could be scrapped for not having any further use, except for one machine to be placed in the Smithsonian as a historical keepsake. (He was right about the one in the Smithsonian.) The idea that computers could do much else beyond math was inconceivable; a computer as a typewriter would have been a joke in 1955 when IBM ventured into strange territory with their 701-series, vacuum-tube-based, electronic computers for the Atomic Energy Commission and the National Security Agency.

Yet even though they believed this was a dead-end market, IBM thought they would learn something from the effort. They did: they learned that the market demanded much more than just the four the secret security agencies promised to buy. By the end of 1955, they had sold nineteen of the 701s, but they lost money on every one of them and the line was closed down. They also learned that the tab-card division—IBM's milk cow—was not happy that this new device might replace tabulating machines. To assuage fears, the IBM 701 used tab cards for input and some memory functions, but that was not enough. The machine was too powerful and was a clear threat to the future of tabulating machines, another reason to close down the line once the government got what they needed.

The second event during that same time frame is less well known, but even more significant. IBM was the prime contractor on the semiautomatic ground environment air defense system that had begun as a crash project in the summer of 1950 after the Soviets exploded their first atomic bomb. As was discussed earlier in chapters 2 and 3, SAGE was the premier interconnected, communicating real-time network, complete with video displays, light pens for marking the screen, and keyboards for computer commands; it even produced the first computer game. This was not the future model for computing that most, including IBM and their advisors, imagined in the 1950s. It was thought a temporary aberration for the duration of the Soviet emergency.

This interactive prototype was not encouraged by IBM, although today we would recognize it as a typical distributed-process telecommunications network, with human interface terminals and data inputs from all over. One of the side effects of SAGE was the training of thousands of IBM engineers in computing to be readied as a sales force for the next generation of machines. Another side effect was the building of the first "wideband" continental microwave network, reaching beyond the Arctic Circle, by its other major contractor, AT&T's Western Electric. A third spin-off was Ken Olsen, the future CEO of Digital Equipment Corporation, who managed SAGE's contractors at MIT's Lincoln Laboratory, and who saw a missing link in the future of computation: the small digital logic processor to do things where it was not cost-effective to harness a gigantic mainframe.

Cultural Lags

Culture can play an important role in resisting change, not only in social environments but in technology as well. People just feel more comfortable doing old things in what looks like a new way, rather than pulling up the roots and using new technologies in totally new ways. This pattern can be seen in the emergence of all the technologies we have discussed in this chapter (including at the dawn of computing) and has become ingrained into the way we approach change. History is useful here, because today we can see how absurdly we behaved toward computers only a generation ago, although in contemporary minds what was done made perfect sense. This should be a lesson for how we are behaving today in similar circumstances of change.

Even though they were moving into an era of general-purpose computing from special-purpose tabulating for their business customers, IBM attempted to keep the future looking as much like the present as possible. Not only was this a successful strategy for capturing business, because it permitted their customers to handle the transition without total shock, but more importantly it appeased the tab-card salesmen who were mounting a rebellion inside IBM against the founder's son. It also locked computing into a rigid model of batch processing, however, when in retrospect this model probably was not necessary.

This is not idle speculation, for the first modern electronic computers were a set of online decryption devices that received signals directly from the airwaves, searched for German grammatical structures, then passed the data on for further machine and human code breaking.[51] That was in 1944—a decade before the 701! Computing started off telecommunicating in real time right from the beginning with "artificial intelligence" for pattern recognition. The next decade of computer development explored a number of models, many of which were similar to the wartime crypto machines. (In fact, some of the first machines owed a significant lineage to these supersecret devices, but it was thirty years before that fact was revealed.)

The SAGE devices were based on a design by Jay Forrester of MIT for a real-time digital computer, the "Whirlwind," which was originally intended for process control and similar applications. Batch processing was not a given, but it did resemble the way most businesses handled rudimentary data on tab card machines.

Technology Push Meets the Regulatory Godzilla

Until the mid-1950s, the digital computer had not advanced much beyond being a powerful tool for the military, and for glitzy exhibitions such as "predicting" the outcome of national elections on television—itself a novel medium. With the 701, and IBM's next version the 702, optimized more for record sorting than algorithm solving, the computer industry became a real business. But first, two more externalities entered the picture, the *first* final judgment in the Western Electric/AT&T antitrust case and the rapid growth of airline traffic, itself a byproduct of government investment in avionics and aeronautics for defense. Beyond the external factors was the rapid development of transistor technology that was overtaking vacuum tubes, which until then had been the basic technology used for switching electrons in the SAGE computers.

AT&T's quasi-legal quasi-monopoly over telephone service had been under different levels of siege since before the Bell patents were affirmed by the Supreme Court in 1888. Here we focus on the antitrust events that impinged on rapidly evolving computer technology and implementation at a critical point: the shift from vacuum tubes to solid-state, all-transistorized machines, which drastically reduced manufacturing and operating costs and permitted significant increases in complexity and functionality.

When IBM entered the computer field, it was dominated by several other companies with more practical experience in computers, and some with equivalent experience in dealing with the business community, including rival tabulating machine manufacturers. Leading the pack was Sperry Rand's Univac division. Univac had absorbed the engineering team that built the Eniac at the University of Pennsylvania in 1943–44, the world's first large *electronic* digital computer, albeit not originally a modern stored-program machine like its contemporaries, Turing's fourteen code-breaking Colossi in Britain. Sperry also made tab-card machines. Control Data Corporation (CDC) was a newcomer along with IBM, but it had a longer, more mysterious history. Earlier, as Electronic Research Associates (ERA), CDC had built a number of very secret machines for U.S. Naval Intelligence even before Eniac and the Colossi. ERA was a code-breaking front organization until they came out of the closet in the early 1950s.

There were a number of such firms in the military complex seeking to establish themselves as computer manufacturers, but the one entity that IBM feared most was AT&T. At the end of World War II, as government's largest single military contractor with its Western Electric subsidiary manufacturing radar and communications gear, AT&T had the manufacturing clout, or so some in IBM believed, to go head on with IBM if computers were truly to become a big business. Both firms had similar models for entering a business—total control of technology and interface appliances, and command of a significant market share, the former under regulatory fiat, the latter by aggressive marketing.

And both firms had continual battles with the antitrust authorities. Both had managed to avoid being broken up, although in 1956 AT&T lost its first significant battle in the original final judgment by agreeing to keep out of unregulated businesses not directly related to telephones. What the regulators did not understand, however, were two significant and related technology imperatives: that computers were going to grow very rapidly as underlying infrastructure for the modern economy, and that computers and communications had already begun to converge as early as the 1950s. The first technological push came from the cold war, the second from the telephone system's constant need for more efficient and more flexible automatic switching plant.

We cannot blame the regulators for missing the full impact of the military's expenditures, for they were well hidden and a perfect example of the law of unintended consequences. The U.S. national security community—and no doubt the Soviets as well—made vast and secret expenditures for computing and communications infrastructure during the height of the cold war. For example, spending on the SAGE air defense system paled in comparison to monies spent on code breaking: the National Security Agency—then an entity so secret that its name did not appear in federal directories and its budget was known to only a few select congressmen—spent over $1 billion per year from 1950 to 1960 on computing devices, computer research and theory, and ancillary technologies such as the development of cheap magnetic tape.

Ignorance of technology and lack of vision, however, were significant problems. This is not much different than the hyperbole we see today about the information highway. The other factor was deliberate obfuscation and dissimulation: AT&T agreed to stay out of unregulated businesses, but its own business needed computing devices to stay abreast of technology to handle increased demand for telecommunications services that simply could not scale without automation. So, the telephone network began to converge with computing in the late 1950s, but to avoid further problems with the regulators, the phone engineers just pretended that telephones were telephones and computers were computers.

The Bell System funded basic research on computing primarily for switching and engineering management of their network. Computers were used to control older electromechanical switches, termed stored-program control (SPC) devices, to avoid making it appear that AT&T was building computers (which they were). Eventually, with digitization of the network, computers switched the bits representing voice and data calls, as they do today, but the pretense was maintained that the telephone network itself was not performing computational functions except incidentally—a mindset which had prevented both policymakers and vendors from seeing how computers were merging with telecom.

The Weltanschauung was reinforced because both government antitrust officials and IBM were not happy with the possibility of AT&T entering computing. Although timesharing had not yet been invented, the model of the telephone network providing computation, or automatic data processing as was the industry rubric then, had several earlier incarnations. In 1940, George Stibitz, a Bell Laboratories mathematician, had demonstrated remote access over a teletype circuit to an electromechanical calculator (built from telephone switching relays) between Bell Laboratories' West Street Manhattan headquarters and the American Mathematical Society's annual meeting being held at Dartmouth College in New Hampshire. Some fifteen years later, the SAGE computers scattered in sites around North America calculated air-traffic patterns with digital radar signals fed from thousands of miles away in order to alert the Strategic Air Defense Command buried in a mountain in Colorado of the possibility of a Soviet air attack over the North Pole. And as the direct-distance dialing system was being built, real-time computation devices were being introduced into the network to route calls, control switches, and automate billing procedures.

Whatever IBM felt capable of building, surely Western Electric could build themselves, perhaps equally as well and as cheaply. Moreover, Western made the best vacuum tubes in the world, and only the best were good enough for computing because, unlike analog radio transmissions and communications amplifiers, computers could not tolerate errors. A bad tube meant the machine was down for repair. IBM bought mostly Western Electric tubes at first. IBM was the prime contractor for SAGE, with Western the major subcontractor. A merger would perhaps have brought computing into the business world faster, and would surely have set the computing/communications model much earlier than it happened; a merger, however, was completely out of the political question.

But computers were not telephones—not yet—so with the 1956 decree, AT&T agreed to stay out of that business, with an implicit understanding that a large fraction of value of the computing device would be supplied by Western's vacuum tubes. Unfortunately for the tube enterprise, Western was hoisted on its own petard, because the transistor that Bell Laboratories had invented some years before came of age just as the Final Judgment was accepted. Before the end of the decade, IBM was producing only transistorized machines, and except for the SPC telephone switches, AT&T missed the formative years of computing in business.

The model of computing was now fixed in the minds of the public policymakers and investors alike as large, stand-alone "mainframe" number crunchers. Fed with data from holes punched in stiff pieces of paper, the machines required very complicated procedures, and very skilled people, to make them compute and display their results on long rolls of fanfold paper. This model lasted almost a human generation before it was dislodged. Yet the model was not necessary, and probably held back the spread of computing to the general public because of the formidable barriers it appeared to present.

The mainframe, batch-process computer not only was inaccessible to ordinary people, but was difficult to use by expert accountants, engineers, statisticians, and scientists who had little direct access to computing. A high priest of data processing had to be consulted, and long delays were accepted before a problem or process was properly programmed and usable results were available. Although farsighted thinkers could envision how powerful machines might become user-friendly (the first personal computers were proposed in 1960), such visions did not impact the policy process.

While other models for computing, especially the merger of computing and communications, were available and change was coming very fast in the underlying technologies, the utilization of infrastructure was very slow in comparison.

As we noted, the SAGE computer/communications model was the complete opposite of the stand-alone, batch-process mainframe, although SAGEs's machines were *truly* large—reaching the size of a two-story building—and its "user" community, the people tracking airplanes, were highly skilled as were the hordes of engineers maintaining the network. Why then, did the rest of the computing world increasingly rely on programmers punching holes in paper, while in the late 1950s, SAGE operators sat in front of cathode ray tube (CRT) display devices with light pens pointing at dots on the screen that actuated computer programs to help detect whether an enemy plane had invaded North American airspace? Why was the human/machine interface model either ignored or rejected as being on the path of the future; a model that some thirty years later we implicitly accept as the only way, for devices as wide-ranging as automated-teller machines or games for the masses to electronic typing, accounting, or management of complex manufacturing process and nuclear magnetic-resonance medical instrumentation? Perhaps the artificial division of computing and communications had something to do with it?

Another unexpected turn of events occurred just as the computer industry began to accept the lucrative role of centralized mainframe host and proprietary peripheral devices. John McCarthy, a student at MIT, was turned down in his request for access to the Whirlwind machines to experiment with some ideas in artificial intelligence, because the machines were both highly classified and too busy with national security matters to let a mere graduate student play with them. McCarthy then solicited funds to build his own machine one that could share processes with humans working on problem asynchronistically—that is, to share in time the machine with different processes—something which he misperceived as the function of Whirlwind. McCarthy did not think he had invented something new, but that he was merely emulating a secret device that he wanted to do work on artificial intelligence theories.

McCarthy's student project led to the formation of Project MAC ("Man and Computers" or Machine-Aided Computation") at MIT in 1962 with the initial support of DoD's new Advanced Research Projects

Agency, General Electric, and AT&T's Bell Laboratories. Under the 1956 consent decree, AT&T could not build computers (although, as we noted, they did as telephone switches), but they could fund research into computation. Again IBM came into the picture, for at this time IBM had begun its design of the famous 360-series mainframes, and timesharing was more of a threat to their model of centralized, stand-alone, tab-card input and output data processing than even the telephone company as "a potential computer utility" could be. The computer utility was raised for the first time as a major public policy issue by timesharing. "Utility" was a bad choice of terminology because it raised early on the fears of the computer industry that it might be regulated, and this resistance kept policymakers from grappling with other critical issues of the emerging information society for almost a generation. Had the future been faced squarely during the 1960s, we may have arrived at an Open Communications Infrastructure earlier.

The rapid advent of timesharing in the late 1960s demonstrated both the future opportunities and pitfalls of communicating computers. With timesharing, terminals were needed in remote locations, and these terminals introduced a fundamental new metaphor into the young field of telecomputing, a method of sending control information for the computing appliances, along with data, in a separate "virtual" channel, facilitating building-block extensibility. These features permitted more than just simple control of physical terminal equipment for it was the first major example of a layered approach to interfacing data-processing equipment and applications, allowing flexible upgrading to more advanced processes and equipment.

Yet layering and flexibility alone do not ensure *compatibility* or *interoperability* with competitive offerings. Indeed, IBM became quite adept at using layered architecture to restrict interoperability—as its proprietary, closed System Network Architecture (SNA) demonstrated by the early 1970s.

IBM approached early telecomputing as they had all other endeavors, from tabulating machines to the first mainframes: they used their market clout to lock in users. It was impossible to manufacture "plug-compatible" devices to work with IBM computers without obtaining a license from IBM for the use of their special EBCDIC character set. The rationale to the users is that nonapproved IBM devices could cause problems. This was not just supposition given that connecting devices is nontrivial, and that a lack of sufficient knowledge or total system responsibility could indeed cause significant problems, as anyone who has struggled to install or upgrade personal computer software knows well. Conversion between ASCII and EBCDIC—today a simple and almost costless function on the corner of a microchip or in software—was expensive in the day of discrete transistors and slow central processing units.

Thinking of the computer as a utility raised issues that neither the large computer manufacturers (primarily IBM), nor the telephone companies (primarily AT&T) wanted to face thirty years ago. But the large user community forced the FCC to face them anyway, although to this day there has been no satisfactory resolution of what is essentially a problem of definition.

Computers confuse definitions because of a mathematical process fundamental to the stored-program logic device that makes no sense to lawyers—recursion. Any digital computer—technically a von Neumann device or a variation of a Turing machine—can imitate any other such machine (albeit at different speeds), and furthermore it can change its own program code, becoming a different machine each time it changes. And moreover, it does this fast—with today's technology, up to some 400 million times per second and tomorrow, *billions* of times per second. This all sounds like nonsense to a regulator who just wants to define various levels of service on a data communications system to determine what is basic and what is enhanced, and what is reasonably competitive and what should be left to the market to decide.

This legacy of confusion has never left us, and never will until the model of what telecommunications is changes to reflect the reality of the paradoxes. Instead of regulating service or utility or usage or traffic, level playing fields are created and mechanisms for interconnection and interfacing are prescribed. This is the essential basis of the Open Communications Infrastructure we have described.

The legacy of the bad definition is now working against the local phone companies. Because data conversion is considered an unregulated "enhanced service"—even though no one in their right mind would use a network to do simple code conversion today—one of the agreements stemming from the AT&T divestiture is that the Bell Operating Companies are forbidden from offering such "enhancements" as part of their basic switched services. Therefore the BOCs cannot offer any data intelligence, including online directory services except through separate subsidiaries. The fine lines that were drawn at the dawn of telecommunicating, have increased in their absurdity over the years as we moved from translating bits representing characters to bits that send full-motion movies over the wires.

Regulators must recognize that what often looks like a good idea, to foster open competition during one time window, rebounds to accomplish just the opposite as technology leaps around the regulatory lag. Technology time windows these days are shorter than regulators' venue on the bench.

Wireless Wars

Wireless telecommunications provides our fourth and final case study of regulatory gridlock. As before, what appears at the outset to be a straightforward embellishment to the existing network turns out to have a profound impact on the regulatory paradigm.

As we have noted, recent developments in microelectronics and computer science have led to wide-ranging improvements in the production of information and communications technologies, with a corresponding impact on the economics of their use.

Mobile communications, before cellular phones, began with early experiments by the Detroit police department with radio communications in the 1920s. These pioneers found that AM transmission from an automobile was too bulky and vacuum tubes too fragile for effective use (Noble 1962). FM transmission proved more robust than AM in the mobile radio environment. Zones of deep interference, or fades, are caused by waves reflected from different surfaces canceling each other. A car rapidly moves through these zones, resulting in degraded transmission quality at low power levels. Early mobile telephone services were based on a centralized broadcasting site, a simple and small frequency range broadcast system that could serve customers to the horizon. These systems were expensive and inefficient, had long waiting lists for subscribers, and required frequent waits to complete calls because of the limited number of available channels.

The Allocation of Cellular Licenses

Early work on a cellular system that could reuse frequencies by broadcasting at low power levels and use several antennas to communicate with users within their region (cells) and switch calls at a central facility was undertaken by AT&T in the 1960s. But the regulatory delays put off field testing for a decade, and eventual commercial operation (in Chicago), until 1982. AT&T, having done the critical basic research on mobile communications, was stymied. The fear was that the AT&T monopoly would grow even more powerful and have a strong anti-competitive impact. Following another decade of technical development led by Motorola, mobile communications technology was finally allowed on the market through two competing service providers in the early 1980s. One service provider would be the regional Bell Operating Company, and the other would be chosen through competitive hearings. Rohlfs, Jackson, and Kelly (1991) analyzed the economic effects of this delay and estimated that it caused a net loss to the economy and telecommunications industry of $86 billion.

It was only in the mid-1990s, at a cost of more than $10 million, that AT&T was able to enter the mobile communications industry which it had pioneered more than three decades earlier.

In the first thirty markets, comparative hearings were held to select the licensees. This proved expensive and time-consuming, so in 1984 a lottery was instituted by the FCC to allocate licenses in the remaining 275 metropolitan service areas and the rural service areas. Consortia were quickly formed in markets 31 through 90, which also received the licenses through comparative hearings, but in markets 91 through 305, lotteries were held to allocate the licenses to a nonwireline service provider (Lee 1988).

Emergence of the Cellular Mobile Industry

Meanwhile capital was invested, cellular systems were constructed, and people bought equipment and subscribed. By June 1990, more than 4 million people had subscribed, more than $5 billion had been invested, 592 systems were in operation and cumulative annual revenues exceeded $2 billion.

The price, size, weight, and power consumption of subscriber equipment were quickly reduced by fierce competition. The first models weighed ten pounds or more and were mounted in car trunks and under seats. Later transportable phones were introduced that could provide three hours of standby time, for example, and were capable of making and receiving phone calls for one and a half hours of full power transmission time. Handheld phones were introduced in 1989 weighing around one pound and consuming less power, although battery size and quality still had important effects on communication time.

Improvement in size and weight has had several implications. The potential subscriber base is increased by making the phones useful in a greater number of situations, and raises the expectations of the consumer. Consumers have also benefited from industry consolidation, which has increased service areas. Licenses have changed hands, through cash payments and company shares, to bring new owners to control many mobile franchises.

Cellular Technology
The performance of the present cellular systems, despite their financial success, still leaves much to be desired. The voice quality is not as good as the terrestrial telephone system; there are places within cities where the reception is bad enough to drop a call in progress; and handoff between two cellular systems is not always successful. Handoff is the signaling by the switching office that a radio transmitter should cease communication with a mobile unit and a new transmitter in an adjacent cell should take over. This is especially difficult when two different cellular systems are involved. People who wish to use their phones in systems outside their home cities, called "roamers," are also now being accommodated more consistently, although significant problems remain. Handling roamers properly requires coordination of information for locating the caller and sending the proper billing information between the systems. Conversations are occasionally crossed, allowing subscribers to hear half of other people's conversations. Privacy can also be more directly infringed on by individuals who use radio scanners to tune into conversations, as is now well known.

A significant problem for the continued growth of mobile communications is the congestion of the present system. As we noted above, the way the cellular system provides for more subscribers is by transmitting at low power and reusing frequencies at several points within the same city.

As long as the power levels are low enough, these transmitters will not interfere unduly with each other or with mobile units outside their cell. By reducing the power and splitting the cells using directional antennas, capacity can be increased. There is, however, a limit to how far this can go. When cells become too small and too close together, the carrier-to-interference ratio becomes small, and noise increases on the phones, degrading quality and possibly causing the dropping of calls. What exacerbates this is usage patterns. For example, people stuck in traffic all use their cellular phones at once and are all within a single cell site. Within a given cell, because of the rapid growth in the number of subscribers and their increasing use of the system, available frequencies become quickly used up. If there are 666 channels in the cellular allocated spectrum, 333 for each carrier (a typical pattern), and a seven-cell pattern used in a city, each cell gets about 40 simultaneous calls, assuming full utilization. With tens of thousand of subscribers in the service area, these channels can be quickly filled at peak times.

This leads to a dilemma on pricing. Charges for using the system are kept relatively high not only to earn money but also to prevent people from talking too long and depriving others of access to the system. But this high pricing discourages people from using the system, either once they subscribe or even when they are considering signing up. But the fact that there are a fixed number of channels available allows the providers to price way above cost, and as a result the shortage of bandwidth translates into nice profits for the carriers. If prices were lower, the increased volume of use would overload the system, degrade the service, discourage other people from using it, and leave them with a bad impression of mobile communications. This situation is a consequence of the present technology and the policy environment, and is at the crux of the pressures for change. Resolving these dilemmas will not be easy. A new block of frequencies was given to the cellular system soon after its introduction, but it was quickly used and only made a small incremental improvement. A change in technology of some kind is needed for mobile communications to expand significantly beyond its present condition, and perhaps even for it to maintain its economic health. This leads directly to the next generation of wireless systems now being planned for personal communications services.

Personal Communications Systems

Several technologies have been developed to overcome limitations on the use of mobile communications. Some are already authorized for use in some markets, others are being tested in the field, and many more are being experimented with and proposed before regulators. Together these systems are known as personal communications systems (PCS). The FCC gave out experimental licenses for many organizations to test them in the field. This is expected by many experts to be potentially a large and lucrative market, given the rapid growth of cellular telephony to date. The spectrum audition process adopted by the FCC for allocation of interactive television broadcasting, and for narrowband and wideband personal communications services in 1994, will be described below, following a brief review of the policy issues addressed as the technology matured.

One of the advantages of the current cellular system was that it was essentially nonproprietary. It emerged from AT&T, whose Bell Laboratories had effectively forsworn patent rights in exchange for the monopoly franchise for U.S. telecommunications, under the 1956 antitrust settlement. When it proposed the cellular system and implemented it, the same method became globally available without large royalties or patent suits crippling competition. Patents, however, often appear to have a chilling effect on the agreement and widespread adoption of standards. At any rate, intellectual property is an important issue to be resolved for diffusion of new infrastructure technologies.

One of the curious features of electronic communications is that the structure of the industry has traditionally been determined largely by the government. Congress, the FCC, and the judicial system determine who can provide what services, under what conditions, and in many cases at what cost. This element of central planning is unusual in the United States economy, especially for an area affecting speech and the press, which is given constitutional protection. As Pool explains (1983), however, there are historical and technical reasons for the institution of these regulations. Antitrust concerns have also played an important role in shaping the industry. The Cable TV Act of 1984, the Communications Act of 1934, the regulations of the FCC, and various acts of Congress have had a decisive influence on the boundaries between service providers, the number of competitors in markets, and how franchises are allocated to provide services. As we have argued throughout this book, the dynamic change in communications technologies has called into question the extant regulatory scheme and industry boundaries.

Another significant issue to be resolved in the PCS market is spectrum allocation. Several issues are raised by the mobile communications industry's claim that it needs more spectrum. One is the allocation of the frequency blocks that the services will use. The second is the awarding of licenses in the particular area where services will be provided. In reallocating a block of spectrum, not only must present holders of the spectrum give up their use, but PCS must justify themselves against every other new service that is competing for frequencies. These include direct broadcast satellites, digital audio broadcast, HDTV broadcast, paging, data, satellite telecommunications, and so on. The existing broadcasters, amateur radio enthusiasts, and other communications services (significantly the government and military users) contest any loss of their spectrum.

An innovative way of determining the area licenses was finally implemented in 1994 and helped decide among different service allocations through spectrum auctioning. Potential licensees bid against each other to receive the right to use the spectrum. At present licenses change hands for millions of dollars on the private market (Jackson 1977). Government-sponsored bidding formalizes the process and captures the funds for public use. The prices paid for licenses would also provide signals about which services were valued more in the marketplace, making allocation decisions possible on the basis of that information as well as all other available information. Spectrum-auction legislation passed in 1993, more than a decade after first being proposed by President Reagan.

A final crucial issue in determining the shape of future mobile communications is the technology itself. The current 30 kilohertz frequency modulated voice channel used by cellular mobile communications systems can be improved in many ways. Switching to digital transmission provides a more robust signal that yields higher voice quality. Adaptive equalization and better modulation methods can be applied to improve the reception of each bit, and degradation of the digital signal can be more easily recovered from than analog noise. Once the transition to digital methods is made, a large variety of techniques can be used to compress the bandwidth necessary to transmit the voice signal. Linear predictive coding and data compression of the digital voice signal allow telephone quality speech to be delivered using as little as 8 kilobits, compared to the 64 kilobits per second used on T1 carriers (Calhoun 1988). In addition, when using digital technology several voice signals can be simultaneously carried on a signal modulated channel, using time-division multiplexing. This method has been adopted in several of the PCS trials. Time-division multiplexing has also been adopted for use in the North American digital cellular and dual-mode analog—digital cellular system. A version of this method has also been used by International Mobile Machines on its digital mobile communications system that it has sold in Mexico and elsewhere.

The alternative to time-division multiplexing, which has been proposed by some, is code-division multiplexing, which uses spread spectrum methods (widely used for military communications for decades) to transmit and receive the signal. This is even more robust than the time-division method and allows a larger number of users to share a frequency band. It is, however, less understood than the time-division method and is considered riskier. Several PCS trials and next-generation cellular phone systems are using the code-division method.

The development of the cellular concept for mobile telephony did not challenge the fundamental paradigm of telecommunications regulation. Telecommunications from cars and trucks was clearly a specialized and unique market, and a unique technology of access made perfect sense.

For a variety of reasons, PCS represents an altogether different case. Although it started initially as a rather innocent extension of a household cordless telephone, as the technology became more sophisticated and the distances of interconnection became greater, the threat to the established industry became clearer.

Poland, Hungary, the Baltic republics and other regions in the former Soviet Union, the Middle East, and Latin America have installed or are now installing wireless, cellular-based technologies as an alternative to a copper-loop system. An examination of the costs and benefits of traditional copper versus advanced wireless designs leads to a most interesting conclusion—in a variety of applications, wireless makes more sense. As the learning for wireless is steeper, those calculations are likely to tip increasingly toward wireless alternatives. What does that signify for the American case?

Until the regulatory process comes to recognize PCS as more than a telephone that can be taken out to the back yard, the implications are constrained. In a nutshell, that is the conundrum this book is intended to address. Wireless access to the local loop offers a unique opportunity to extract a special public benefit from existing technologies. Wireless access requires a paradigm shift. The balance of the public interest between regulated monopoly common carriage and competitive local service provision becomes, for the first time in a century, a serious question for research.

When the wireless telephone itself first appeared, few could imagine its ultimate growth and impact, and certainly not the course its history would take. Change in the existing structure of the industry will undoubtedly be resisted by entrenched interests. It is a complicated case because most companies have divisions drawing substantial profit from wireline, wireless, and other enhanced services. The net result, however, is most interesting indeed. The connection between the entrenched interests and the regulatory process may finally have been broken.

A Pattern Emerges

We have just completed an abbreviated and self-conscious selective history of the evolution of the communications industry in the United States from the turn of the century. The dominating metaphors involve wars and warriors. We characterize these developments as an ongoing battle as the established players attempt to ward off a variety of potential competitors. These players use their two primary strategic advantages— their market power and their ability to manipulate the regulatory structure—with some success. The challengers rely on new ideas and new technologies.

We review a history but not as historians whose stock-in-trade is capturing the historically unique. On the contrary, as social scientists we are less interested in the historical specifics than we are the recurring pattern of industrial competition and regulatory change. Our purpose is to diagnose the pathologies of this process and recommend a policy physic, which we will address in the chapter 6.

We are convinced that a pattern is evident in the network wars of the last century and that as we enter a particularly intense period of competition in the buildup and shakedown of the American National Information Infrastructure, it is prudent to examine those patterns closely.

We identify nine recurring patterns that appear to dominate this history. With the exception of the first one, each pattern implies a pathology, a problematic aspect of the battles among vested interests in the communications domain that, in our view, does not serve the public interest. We are not opposed to network wars. Unlike their military counterpart, they require fewer resources and engender much less human suffering. They are the natural and healthy outgrowth of a technically and structurally dynamic communications industry. But we believe the rules of engagement and the customs of battle can be fruitfully assessed and reformed so that the conflict is as fair and productive as possible.

Technology push   The pattern repeats itself in many domains, successive waves of conquering armies, new technologies pitted against older established ones. We are cautious to avoid an unthinking technological determinism, yet over and over again in the history of network wars, the disruptive force is a technological one. Technical invention is the first mover that forces the redeployment of resources and the questions of established patterns of doing business.

Political gridlock   It is frequently said in policy circles that it is easier to stop a policy initiative or new piece of legislation than to get one passed. Special chemistry is necessary to array forces and line up support sufficient to hammer out a compromise or overwhelm the opposition in full victory. Such chemistry is rare. As a result, most politico-economic power battles between and within industrial sectors are indeed battles of titans and the resultant high-power gridlock is difficult to resolve.

Cultural lag   In addition to changing players in these battles for economic eminence, we confront changing ground rules for how the battle is to be fought. In military history it is often observed that the generals too often strategize on the basis of lessons learned in the last war rather than the one they are fighting. The same phenomenon is evident in the network wars. CEOs who rose through the ranks during AT&T's mandated telecommunications monopoly developed skills in the tariff game, but not necessarily the kinds of skills for management agility necessary in a competitive telecommunications environment. And from our review of telecommunications, broadcasting, and computers we would conclude that this cultural lag extends to the regulators as well. Regulators come to recognize certain patterns of "abuse" based on a technological paradigm of communications derived from case law a decade or two old. In times of rapid change, however, such paradigms prove to be outdated.

Narrow horizons   It is sometimes remarked that when American business executives turn their attention to long-term strategic planning, they are referring to the next two quarterly reports. The incentive structures of the modern American economy, it is frequently observed, do not provide many rewards for executives who might otherwise gravitate toward longer-term issues of industrial structure or technology. This applies as well to regulators, policymakers, lobbyists, and analysts. One is drawn to focus on the heat of the battle rather than on the direction of the war. Players stake their flags in support of or in opposition to a bill, a proposed regulatory change, or the introduction of a new competitive service. Once the battles get started, the soldiers lose track of what they are fighting for and how it might possibly relate to their companies' or their industries' long-term strategic goals.

Dissimulation   The most artful combatants press the case for their vested interest while appearing to champion the cause of fairness, technical progress, and the public good. To admit to self-interest publicly weakens one's case. As a result, dissimulation and obfuscation are rampant. What would appear on the surface to be a technical debate over optimal image quality or most efficient digital transmission is really a battle over who gets patent rights, competitive advantage, and market share. Such strategic pretense is understandable, but it clouds and distorts the process to the disservice of the players, the policymakers, and the public alike.

The asymmetry of established player and newcomer   Inertia favors the established players. They have ample resources to preempt, co-opt, and litigate if significantly threatened. Players prefer a known competitor than a new one whose strengths and weakness are not yet clear. Of course in the history we have recounted, the numerous Davids do well against the Goliaths. But in the real world the process is long and arduous.

The asymmetry of vendor and customer   Even as we move toward more competitive provision of communications services, the system operator is in a position to know a lot more about the costs and capacities of the network than even a technically competent high-capacity user. It is a natural asymmetry. The customer has his own concerns to worry about, and thus stands at a disadvantage in negotiation and conflict.

The asymmetry of regulator and industry   The regulators have the power, but the regulated, it would seem, have everything else—financial resources, technical and financial knowledge, the capacity to appeal to the public and the political elites. Typically a regulatory docket is initiated by industrial players, the arguments are framed by their representatives based on financial or technical data they collect and interpret. As the rate and complexity of change increase, the asymmetry is even more pronounced.

Standards incentives   From the point of view of the system's user, the more universal and interoperable the technical standards of the network, the better. From the point of view of a network provider, interconnection is a dangerous business—an open invitation to competition. The strategic incentive for established providers is to resist interconnection and standardization in an attempt to control as much of the communications flow as possible. When new generations of technology are incompatible with their predecessors, hardware manufacturers are enthusiastic—it removes the last generation as a meaningful competitor. Such incentives work only up to a point. It is a balancing act. How much incompatibility can vendors persuade customers to buy? The second-order effect is that customers see thorough the interoperability tactics and will make purchasing decisions based on standardization.

6

Cutting the Knot

Several thousand miles east of Anatolia where Alexander cut the Gordian Knot, is the scene of another historic and symbolic conflict, this one even more ancient. The scene is the Gobi Desert in the Jurassic era. Two dinosaurs locked in combat apparently did not even notice the approaching sandstorm that buried them both. Their fossils, with their jaws still clasped around each other's neckbones, were not discovered until the summer of 1994. What a wonderful symbol of gridlock, two dinosaurs so intent on mutual destruction that they were totally unaware of the impending storm. One might be tempted to draw an analogy to the present-day battles of communications industry moguls. Given our faith in evolution, we should try to resist. But the temptation is strong.

The sandstorm of technical change, we argue, requires a paradigm shift for federal communications policy critical to survival. Since the era of the robber barons, the technical character of telegraphy, telephony and broadcasting required that each be regulated as a public-service monopoly. These are large-scale enterprises, centrally and hierarchically controlled, and heavily regulated. We posit that in time this will come to be known as the Jurassic era of communications populated with large firms, quaint technologies, and complex regulatory policies—dinosaurs all. The digital revolution can best be served by open, undistorted, healthy economic and technical competition. As we reviewed the literature in this field, we found near universal acknowledgment that in a vaguely defined, distant future competitive provision will be the norm, but until then there must be reams of new legislation and armies of regulators to manage the transition. We disagree. It is best to proceed quickly, because it will prove impossible to untie the knot one cord at a time. We have termed our proposal Open Communication Infrastructure.

The first paragraph of the 1934 Communications Act charged the FCC with regulating "commerce in communication by wire and radio so as to make available, so far as possible, to all the people of the United States a rapid, efficient, Nation-wide and world-wide wire and radio communication service." It represents a most interesting choice of words. It would seem to envision a single, integrated, open communications infrastructure via spectrum and wire to serve "the public interest, convenience and necessity."

Of course, Titles II and III of the act reinforced the distinction between common carriage, wireline telephony, and radio broadcasting—a legacy of the Cullom (Interstate Commerce) Act of 1887, the Mann-Elkins Act of 1910 and the Federal Radio Acts of 1912 and 1927. But by all accounts, the 1934 act, as reflected in these words, was designed to bring together the existing regulatory machinery from both traditions (from the Interstate Commerce Commission and the Federal Radio Commission) into a new and unified Federal Communications Commission. President Franklin Roosevelt's letter in support of this legislation emphasized the importance of bringing together the two domains of communications. The president's administration study group, chaired by the Secretary of Commerce, also supported this approach and expressed concern about a "collection of communications agencies not working in accordance with any national plan" (Paglin 1989).

The passage of the 1934 act and the establishment of the FCC were surprisingly uncontroversial (Robinson 1989). They were seen by some as straightforward administrative conveniences. Most of the commercial interests that now monitor and lobby the commission did not exist in 1934. The broadcasters were a diverse lot, who from 1926 had worked hard to get the federal government to play an effective role of traffic cop in systematizing broadcast frequencies and reducing interference among stations (Barnouw 1966). AT&T under President Theodore Vail welcomed strong federal regulation and protection in accordance with the Kingsbury Commitment of 1913 (Brock 1981). Since then, industry and the commission's own bureaucratic structure crystallized into three separate and independent areas: common carriage, private radio, and mass media. What an irony: The keystone of the 1934 act, the development of an integrated "rapid, efficient, Nation-wide and world-wide wire and radio communication service" seems to have gotten lost in the bureaucratic shuffle.

More recent attempts to rethink these premises, or to rewrite the 1934 act in light of all that has changed since the 1930s, however, have provoked a firestorm of opposition. The antagonism stems from the many vested interests that have established viable businesses under the existing rules and fear what rethinking or rewriting might entail (Haight 1979; McHugh 1983; Weinberg 1983). The divisions are clear-cut, the incentives to defend turf are powerful, and the political and financial resources of the stakeholders inspire awe even among experienced Washington observers.

In 1934, conditions were ideal for developing a new initiative. The economy was gradually rebuilding. The New Deal was in full swing. The act was a relatively small element in a larger process of political change and reformation including the Securities Act of 1933, the Securities Exchange Act of 1934, the National Labor Relations Act of 1935, and the Civil Aeronautics Act of 1938. The key industrial players in communications had calculated their gains and losses and had become comfortable that systematic federal regulation was the safest bet. And, most important, the division of turf had been prenegotiated by the dominant firms of the day as a result of the General Electric/RCA/Westinghouse/AT&T cross-licensing cartel arrangements of 1919–1921. RCA did the broadcasting and was in large measure owned by the others, GE and Westinghouse manufactured equipment, and AT&T carried network traffic over its long lines but otherwise got out of the radio business, although it would later gain control of sound recording for film.

The pressures for change and reform of the organization and regulation of the information/communications industry on the eve of the twenty-first century are as great as they were in 1934, perhaps greater. But forces resistant to change are much stronger as well. The complexity of the relationships among the innumerable players, when information has become the currency of business, and the intricacy of international corporate alliances confound the problem. The spectacular backroom deal of the 1920s cross-licensing cartel was hardly forged with the public interest foremost in mind and is certainly not a promising model for industry self-reform in the 1990s. But something of similar magnitude is needed if the public interest is to be served at this critical turning point in communications. The Communications Act of 1996 falls woefully short of what is needed to overcome gridlock.

Over the years a variety of techniques has been used to break political and industrial gridlock. For example, J. P. Morgan invited the feuding magnates onto his boat in 1885 in Long Island Sound, and refused to let them off until they settled the railroad/telegraph (nineteenth-century convergence) wars, which resulted in the formation of AT&T to counterbalance Western Union—something that was hardly noticed at the time. But within a short time the federal government had to act to keep the empires from destroying each other and further exploiting their customers. Infrastructure was too important to leave in the hands of the robber barons.[53]

This pattern of deal making, and deal failure, repeated itself every quarter century or so, each time yielding tighter governmental regulation than before; the deals seldom held but the deal-making process ill serves the public weal.

Who will be the new J. P. Morgan? Who has a big enough boat to hold the barons of Hollywood; computing hardware, software, and services; consumer electronics; telecommunications; publishing; cable; and broadcasting? Will representatives of health, defense, education, and government services be permitted on board? Will anyone speak for the public interest? And how long with this deal last? We believe the time has come for a new approach, a new deal: OCI.

A New Policy Architecture: Open Communications Infrastructure

A new policy approach is needed. Policymakers must set clear-cut, unambiguous objectives about the nature of the national communications infrastructure, then let the entrepreneurs hammer out the details among themselves about how best to achieve these objectives. Individuals and groups would be empowered to develop and use the services appropriate to their needs and interests.

The new policy architecture of OCI would offer fully competitive provision of all local and national communications services. All regulatory distinctions would be erased between wireline and wireless communications services, narrowband and broadband, broadcast and switched communications services, and content and conduit. A unified policy structure would eliminate the distortions resulting from the unequal regulatory treatment of different infrastructure and service providers.

This view explicitly rejects the Japanese tradition of MITI-style planning and the European corporatist system. Direct government involvement in planning and building the network itself is not part of the American tradition of government-business relations. But government leadership and support for private development of infrastructure is very much in the American tradition. In our view, continuing this tradition with a coherent set of policy goals is clearly the best-case scenario for the future.

A National Goal

The enhancement of commerce in communication by wire and radio so as to make available, so far as possible, to all the people of the United States a rapid, efficient, Nation-wide and world-wide wire and radio communication service.
Communications Act of 1934

This goal should sound familiar, and it is as valid now as it was in 1934.

The Means to the Goal

We recommend a proactive national policy to make the provision of communications and information services competitive. We call this approach Open Communications Infrastructure. It is premised on the belief that if service provision is truly competitive, content and ownership regulations in broadcasting and common-carriage rate regulations in telecommunications will no longer be necessary. Once the policy is put forward with clarity and conviction by the administration, Congress, regulators, and the courts, the response from industry will be positive, creative, and surprisingly swift. The elements of an Open Communications Infrastructure are:

1 **Open architecture** Future digital networks will transport electrons and photons representing voice, text, data, and images with equal facility. Some messages will require extensive spectrum or wireline bandwidth, others will not; both may provide two-way interconnectivity. The heart of this approach is flexibility, or designing a system that facilitates interconnection among different systems and services at present and as they develop over time. By emphasizing interconnectivity and interoperability, incentives in innovation shift from trying to protect licensing revenues by means of a proprietary technical communications standard or inherited spectrum allocation, to the economies of scale in efficient manufacturing and service provision. The positive externalities realized by opening the network to everyone on the network as potential sources of innovation are tremendous, just as the growth of the Internet has demonstrated. Open architecture enables users to become producers, and service providers to reach a wider marketplace.

In practice, an open architecture is always a relative term: open to certain forms of interconnection and certain services, but not to others. Nonetheless, open architecture is a prerequisite for a competitive information marketplace, because without flexible interconnection, suppliers and customers are limited in the range of applications and functions that can be performed through advanced networks.

Implementing an open architecture requires detailed technical planning best performed by technical professionals, not by government. Government can help by enunciating clear goals, and being responsive when firms are the subject of complaints that they are violating the principles of open architecture design. The case of the set-top box, or interface to broadband cable and video programming, is an example of the interface issues that government must still be prepared to monitor, even when practicing regulatory forbearance in an Open Communications Infrastructure. While staying out of the details of service design and provision, ensuring that services offered to the public employ an open architecture design wherever possible is an appropriate role for government, whether it is the courts, the Department of Justice, the National Telecommunications and Information Administration, the Federal Communications Commission, the Federal Trade Commission, or state public utility commissions that is most concerned with a particular interface issue.

2 **Open access** In spite of the deregulatory rhetoric in Washington, the dominant model of communications economics presumes that limited spectrum and inefficiencies of the competitive provision of wireline communications services require legal barriers to competitive entry and regulatory oversight of monopolists. This premise is outmoded. Open access means that the presumption should be in favor of interconnection and the elimination of regulatory barriers, not the opposite. Public policy should be oriented toward encouraging open access as a critical goal.

Open access is reinforced by the increasing diversity of potential competitive paths into the home. For example, wireline access via cable networks for data communications or telephony may be extremely competitive with the telephone network. In the United Kingdom, within a few years of service introduction, 20 percent to 30 percent of homes passed subscribed to phone-service offerings by their cable providers. High data rate Internet access via cable companies also appears to be an attractive potential service offering for residential and business customers (Gillett 1995). On the other hand, telephone companies are looking to hybrid fiber/coax networks, which would enable them to compete with cable industry television services.

The set-top boxes of the future, or home hubs, whether they are built into televisions, computers, telephones, or indeed are sold as set-top boxes, may be the new interface point that must be kept open for electronic commerce to flourish in the future. If neither telephone nor cable firms have an assured monopoly over service provision, the threat of competitive entry into contestable markets should be sufficient incentive to keep the competitors honest, and charge prices related to marginal costs. When they are not, and abuse of a monopoly position can be documented, traditional antitrust and fair-trade laws should be able to do much of the rest, with no need for regulators to get involved in detailed-rate or standards-setting proceedings.

In addition, wireless access to the local loop may be fully competitive with wireline provision. Wireless access to the network has other properties, especially in cellular and microcellular applications, such as personal communications services. The critical property is the lack of steep economies of scale in service provision. To offer cellular service one can start at a small scale. For fewer customers, fewer cells are required. As service demand increases, the number of cells expands gradually and efficiently, without requiring additional bandwidth. Another property of course is mobility and the flexibility of personalized communications service. In this way, the promise of an Open Communications Infrastructure may be to a certain extent an unanticipated artifact of cellular technology. Wireless access to the local loop makes open competition in provision of local exchange service a meaningful prospect, not in some distant future, but here and now. The sale of spectrum for new wireless services creates new entrants with strong incentives to compete for customers with the incumbent wireline network operators. In this case again, striving for open access and interconnection becomes a predominant policy objective as most regulatory mechanisms wither away.

3   Universal access   The deregulation of communications raises fears that less-privileged and more remote communities will be deprived of access. The fear is that competitive provision will lead to the decline of universal service (Hadden 1993). Critics, for example, point to the deregulation of airlines and the resultant changes in the cost and quantity of service to remote areas. The economies of scale in providing airplanes and airports, however, are still considerable. Moreover, the variety of airline services and their cost has widened, so that from an economic viewpoint it does not appear that access to airline services has declined in cost, quality, or availability. Wireless works the other way. In fact, it is much more efficient for low-density applications in rural areas. Furthermore, direct satellite provision offers additional avenues for service extension and system redundancy. Universal access is a natural component of an Open Communications Infrastructure.

Universal access, rather than universal service, is a more accurate term to describe the traditional goal of public policy for telecommunications services. In India, universal service could be achieved with one telephone per village. In the United States in the early twentieth century, universal service meant anyone who could afford a telephone should have nondiscriminatory access to telephone services. Over time with the advance of technology and increased national wealth, the definition of universal service expanded to mean that every home should have access to a telephone. By the time of the 1992 Cable Act, universal service appeared to be defined by basic access to television as well as telephone services. In the future, universal access to the Internet or the National Information Infrastructure will be the measure of universal service, or again as we prefer to more accurately call it, universal access.

No matter how universal access is defined, it is clear that it is a universal and inherently political term. That is, measuring whether universal access is achieved begs the questions of access to what, by whom, at what price, and for what purpose. Rather than attempt to envision and codify detailed rules for all possible permutations of universal access, or universal service if one prefers that term, we prefer to leave this to future public policy debates and actions by industry and government.

4 **Flexible access** An interconnected and interoperable network of telecommunications, broadcasting, and electronic publishing breaks down traditional distinctions. In OCI, one can make a call on the cable, send a fax over the radio, and watch television from the telephone line. The bandwidth will be adjusted according to the demands of the user and the character of the communications. Such distinctions need not be legislated; a truly competitive market will provide more diversity than the law could imagine. But the fundamental flexibility and interoperability in communications architecture is not an inevitable outgrowth of current market dynamics. The role of public policy in defining the nature of the market is critical.

Flexibility in design and provision in an open market for communications services is an important but elusive goal. Precise regulatory measures of flexible access are hard to imagine; perhaps it is a bit like pornography—we know it when we see it. While a precise definition may be unattainable, flexible access is an important goal to ensure that services are available in a variety of configurations depending on individual needs and interests.

Problems of Implementation

This book has attempted to present Open Communications Infrastructure as an attractive, persuasive, and coherent concept to guide a strategic and forward-looking reformulation of American communications policy. We feel strongly that the fundamental approach is sound and is especially well suited to the American situation. The issues addressed here are currently working their way into the central forums of communications policy in the United States. We conclude with some of the comments on implementation. The current debate over deregulatory strategies is highly contentious and complex. Basically we argue that the details of implementation can be derived from general principles. If one scratches the surface of the current debate over legislation and regulatory reform one quickly encounters the self-interested rhetoric of one group of stakeholders or another. They fear their competitors will continue in the old ways to stifle competition and somehow derive unfair advantage. They want safeguards and all sorts of new rules to govern the transition. It is a reasonable strategy for a communications company to pursue, but it makes lousy public policy. We argue that, in large measure, the old strategies simply will not work in the new environment and the safeguards game serves primarily to delay the movement to competitive provision.

The question of time frame   As this is being written, a new communications legislation has just been passed. Each of the previous rounds introduced with a flurry of optimism ended in legislative gridlock. The rhetorical first paragraphs of these bills, as usual, talk of competitive provision but most of the paragraphs that follow speak to the details of relative advantage during a long and presumably arduous transition. One new rhetorical flourish this round, however, attracts our attention— the phrase: *date certain.* Such a notion does resonate with the OCI vision. Pick a date in the not-too-distant future and instead of battling over definitions of competitive thresholds, and indexing deregulation to those complex calculations, simply declare communications deregulated as of that date. Let the stakeholders beware. Watch the investment of time and energy shift from lobbying to the development of new products and services. Such rhetoric, however, does not sit well with the regulatory set. They cling to the old paradigm, conjuring up scenarios of inequity and unfair advantage. They prefer date uncertain, it is so familiar and reassuring. The final legislative draft replaced date-certain language with a complex nine-point checklist which virtually guarantees further legal wrangling and delay.

Inertia and the power of dominant players   Initially, the newspaper industry resisted the radio industry as an upstart competitor, but the newspaper publishers were smart. When radio showed promise, newspaper publishers bought stations, to their significant financial benefit for decades. The same pattern played itself out as cable television came to challenge the established broadcasters. If Open Communications Infrastructure evolves as we propose, the currently dominant players will draw on their two primary strengths, their expertise and their existing plant, for competitive advantage. In the transition from monopoly to competitive long-distance telephony, due care was taken to protect the balance between the new players and the dominant carrier until the new marketplace stabilized. Caution is appropriate, of course, but we are not convinced the dominant carrier-new carrier model is apt. Perhaps more appropriate is a model of dominant players from different industrial sectors now in increasingly close competition with each other as boundaries blur. It's dinosaurs against dinosaurs. Perhaps we need not fret quite so much over survival of the weak.

The powerful resistance of vested interests   Some firms will no doubt calculate the relative advantages of muddling through with the current system of mixed regulations by comparing the known economics with the likely market dynamics of OCI, and these firms may conclude that the former is more likely than the latter to serve their economic needs for some time. They will make their views widely known in public and in private. Albert Hirschman's recent historical study of the rhetoric of conservative reaction to reform movements (1991) lets one predict the form these criticisms will take. If history is any guide, these small firms will also energetically proclaim that the public interest itself is ill served by movement toward open and competitive communications. They will likely argue the proposal is perverse (that the actual effects of OCI policies will be other than what has been predicted), that it is futile (that common carriage and content regulation will continue to be necessary), and that OCI puts the communications infrastructure in jeopardy. (Important traditions and values will be unnecessarily cast aside.)

"Who is us?"   The language of this book emphasizes the importance of meaningful American participation in the design, development, and management of an advanced national communications infrastructure in the United States. As Robert Reich's now famous query implies, the question of how to define meaningful participation becomes increasingly complex as firms and economic linkages span national boundaries. Any nation's communications infrastructure is an important element of national sovereignty and political integration. The 1934 act unambiguously and appropriately limits foreign ownership of telecommunications and broadcasting infrastructure. The importance of national control over communications infrastructure is of course highlighted in times of civil emergency and war. It is an appropriate focus of legislative attention, but it is also a potential weapon for those who would prefer protectionism to meaningful global competition. Times of war and civil emergency are by definition exceptional. We would prefer a legal structure, like in most other areas of national infrastructure, that comes into play only when necessary in times of crisis. Regulation of foreign direct investment is too blunt a regulatory tool to achieve the policy objectives in question here.

Limited spectrum   Is there enough spectrum to make competitive access viable? Our analysis indicates that there is. The key question of course is wireless access over what geographic distance. Microcell technologies permit dramatic efficiencies through geographic reuse of spectrum. Given the flexible, geodesic structure of OCI architecture, as spectrum for wireless access gets crowded, the incentives rise to move the wireline node closer to where the demand for communications resides and, in turn, reduce spectrum demands. Providing an alternative communications path in wireless access takes place through hand-offs in fractions of a second. Providing alternative wireline nodes takes weeks and months, but it can be speeded up when the market demands.[54]

Content regulation   Principles of press freedom and common carriage have guided a relatively unobtrusive tradition of content regulation thus far in the United States. The convergence of media offers a window of opportunity for those uneasy with this lack of public control. The targets are familiar: unpopular political views, explicit sexuality, and violence. Although one can expect numerous attempts to identify and control the flow of information for both good and bad reasons, the distributed network of networks, the use of encryption by parties who wish to communicate with each other, and the globalization of the communications process will frustrate the impulse to control.

Technical standards and mandated interconnection The fear also persists that without central control, the evolving networks will not choose to connect or will not be able to connect to one another because of technical incompatibilities. Although the players may continue to dream of de facto monopolies and de facto proprietary technical standards, neither is at all likely to prevail in the new environment. As they continue to jockey for dominance, and threaten incompatibility, the user community and the policy analysts wonder out loud whether it might be necessary to return to the old system of legally mandated standards and interconnection. Although a network of networks is chaotic and confusing, it relies on the tipping phenomenon of network externality economics to maintain interconnection rather than legal mandate. Neither system is without cost, neither is free of inefficiency. But to mix the two is likely to generate the worst of both worlds

The Role of Congress and the Administration
The 1934 act was a joint product of the efforts of the Roosevelt administration and the seventy-third Congress. If powerful individuals on either side had strong objections, the bill would have been scuttled, and industry would have been forced to make do with the patchwork arrangement of existing legislation. The cooperation was fortuitous. Just as fortuitous was the cooperation of the Clinton administration and the Republican-majority one hundred and fourth Congress which crafted the 1996 act. Unfortunately the new act falls so far short of what is needed that further legislation will be needed soon. The issue of telecommunications reform was not "solved" by the 1996 act; its need was only made more apparent.

Given the breadth of the issues involved here, a cooperative spirit is again of great importance. The notion of an integrated and proactive communications policy with important benefits to industrial productivity and employment is likely to attract positive attention in Congress. The reliance on industrial initiative and competition, rather than bureaucratic industrial policy, is likely to resonate well with the current political mood.

The Role of the Existing Regulatory Agencies

It could be argued that putting the FCC in charge of a proactive and truly deregulatory initiative would only serve to delay if not derail the desired end state. The commission, however, has the expertise and experience to manage a successful transition. After all, the Civil Aeronautics Board successfully plotted its own demise through a policy of deregulation. Perhaps an entirely new agency (as was the case in 1934) would be needed, specifically structured for the job at hand. The Gordian Knot argument, however, insists that whatever the organization, the vision must be clear. OCI requires no permanent federal communications bureaucracy. The dozens of new regulatory proceedings mandated by the 1996 act generally serve to perpetuate the old model of telecommunications as a regulated industry, and threaten to extend federal involvement into micromanagement of content production, data networking, and information service provision.

In 1995, Assistant Attorney General for Antitrust Anne Bingaman, with the support of Attorney General Janet Reno, allowed Ameritech to enter the long-distance market in selected cities. The test case required evidence of actual local telephone competition and the provision of interconnection and market access to competitive local telephone providers, presaging the restrictions in Title I of the 1996 Telecommunications Act. We see no reason to restrict entry by means of legalistic and bureaucratic checklists. Traditional antitrust mechanisms can serve to keep the incumbent firms honest. Let's call it the Bingaman Commitment. But we must extend from this cautious half-step, and remove all restrictions on entry. The motivating factor is the fear of being left behind as other firms venture into new markets. The balancing factors that prevent unfair manipulation of a player's current stake (including networks and customer base) need not be a new set of bureaucratic thresholds that attempt to measure "meaningful competition." We believe two existing factors are sufficient. The first is simply the threat, now the reality, that competitive entry protects the consumer's interest better than the regulatory process ever could. The second is to rely on the existing threat of antitrust litigation. The recent antitrust investigations of AT&T and IBM are fresh in the minds of the executive players and lest they forget, the Microsoft antitrust inquiry lurches in and out of the news as a frequent reminder.

We might note in passing that GTE, while providing local telephone service to 20 percent of the U.S. population, was never legally restricted from the long-distance business. After investing in (and subsequently divesting) Sprint, GTE's corporate strategy focused on cellular telephony and local service and now strategic partnerships with long-distance carriers. As this example shows, the removal of restrictions on market access does not automatically mean a firm will invest in that new field, only that it will have the opportunity to choose its investment strategy based on business criteria, rather than unneeded regulatory barriers.

The phenomenal growth of the Internet illustrates that there is a large economic benefit for users and service providers to be gained from interoperability and interconnection. The ethos of sharing and cooperation that is basic to the Internet culture could spread throughout an interconnected Open Communications Infrastructure. The Kingsbury Commitment of 1913 provided for the growth of the telephone network and the extension of universal service with a pricing scheme that was considered fair in its time. The Bingaman commitment could provide the path toward universal, open access to the National Information Infrastructure.

As with the entry of local telephone firms into the long-distance business, there is no reason cable television firms should be a priori excluded from residential and business telephone service markets. Whether the service is via wires (or fiber-optic cable) or reaches homes via wireless technologies should make no difference from a regulatory point of view. Spectrum auctions and the growth of unlicensed spectrum services are harbingers of the economic benefits for users, firms, and taxpayers that can be gained from a more flexible approach to spectrum use.

The generic properties of the new networks are inherently unfriendly to the instinct to create monopolies, build hierarchies, and centralize control. The locus of control migrates from systems operators to users. The movement from single-purpose to multipurpose networks enhances the capacity of competition. The century-old traditions of commoncarriage and public-trustee regulation become unnecessary burdens as both regulators and systems operators struggle to deny that their evolved expertise at regulatory gamesmanship is no longer relevant.

Furthermore, the likely compromise of a half-regulated, half-competitive system of communications networks may actually be worse than either by itself. In our view, such proposals for reform threaten to build in the worst of both worlds as the economic distortions of the regulatory process and the regulatory distortion of the competitive process diminish the effectiveness of both while dramatically slowing down the pace and incentives for technical progress. They are trying laboriously and mightily to untie the Gordian Knot one strand at a time. It will never work.

The policy gridlock must be cut dramatically by means of a clear and distinct policy initiative. Such an initiative posits that regulators are ill equipped to micromanage such fast-moving technical developments. It posits as well that industrialists who for the most part have enjoyed a century and a half of near-monopolistic profitability will do everything they can to recapture and monopolize the new network structure. Some media executives seem to perceive it as something akin to a natural right, others perhaps simply as sound business practice. In any case, a laissez-faire approach of leaving the transition to "market forces" alone is a recipe for disaster.

A Look to the Future

As we review the current state of affairs, it strikes us as self-evident that the dramatic changes in communications technology challenge the accumulated regulatory structure so significantly that a complete reevaluation of the appropriate role of government in communications policy is necessary. The time is right for a bold initiative. The Communications Act of 1996, and before that the National Information Infrastructure initiative of the Clinton administration, are small steps toward (and away from, at the same time) an Open Communications Infrastructure.

We now have the opportunity to design a new electronic and optical network that blurs the distinctions between mass and interpersonal communications and between one-way and two-way communications. The phrase "new media" is misleadingly plural. Most reviews of the field list the growing array of electronic devices identifying the special properties of each from direct-broadcast satellites, personal computers, digital, high-definition and interactive television, videotext and teletext, electronic mail and high-speed computer networks, to a variety of enhanced services for an expanding digital telephone network. In the end, however, the new media are one—a single, high-capacity, digital network of networks that bridges what we now know as the separate domains of computing, telephony, broadcasting, motion pictures, and publishing.

Each of these industrial sectors is currently highly profitable in its own sphere. The market boundaries between these sectors are based on a series of evolved social conventions for the repertoire of media technologies appropriate for each category of human communication. A single integrated electronic system for high-quality video, audio and printed output makes such artificial barriers less meaningful. As a result, every corporation in any of these fields will soon face three or four times the number of determined and well-financed competitors for their business, a prospect about as welcome as an invasion by the Vandals and the Visigoths.

The American political tradition in such matters is laissez-faire. The concept of even a broadly focused reformulation of communications policy for the information age is most difficult, but rethinking regulatory fundamentals is clearly critical now.

Some of our colleagues, whose expertise is in engineering and advanced electronics, advise us to relax, sit back, and enjoy the battle of the Titans. After the battle rages on and their resources are exhausted, the technology will go its own way, regardless. An open communications infrastructure will emerge, Phoenix-like-the result of technological magic. Other colleagues, whose expertise is in economics, advise us to watch closely and learn from the battle. It is the invisible hand making everything right. If a convergent and integrated network does not evolve, it should not so evolve; the gods of economics know better and think otherwise.

We remain concerned. Our skepticism of magic and alchemy persists. Our belief is that public policy will play a role in the evolution of the social and economic electronic infrastructure for the next century. The growth of the Internet is an encouraging sign of what may come, but as we write the net is being besieged by the forces of reaction—in the Congress, the Department of Justice, currently dominant industries, and among an ill-informed public. Our colleagues in Europe and Japan have recognized the need for a fresh, coherent approach to national communications policy. It is now time for an American initiative, extending far beyond the censorious regulatory morass of the 1996 Communications Act.

In putting forth the model of Open Communications Infrastructure as an American alternative, we believe we are playing to the strengths of American business and the American political tradition. We feel the choice is to wait hopefully for it to happen by itself, if it does happen, or to make it happen. Technology can make it possible, but policy can either thwart or enable it. The central goals are clear:

• Fully competitive provision of all local and national communications services.
• Lifting all distinctions between:
· wireline and wireless communications services;
· narrowband and broadband;
· broadcast and switched-communications services;
· content and conduit.

The development of an intelligent and forward-looking American communications policy for the next century will require a great deal of research and analysis. Our purpose here is to put a proposal on the table for further discussion. This book develops the case for a proactive approach to revitalizing the American communications infrastructure as forcefully as our resources and capacities permit. It is the product of independent academic research. It is not the final word, but it is, we hope, the beginning of a debate.

Notes

1. Telegraph networks, being composed of devices that sent pulses instead of waves, had a somewhat easier technological task for repeating signals, so telegraphs could be sent virtually round-the-world, while telephone signals, until the invention of the vacuum tube amplifier, were restricted at most to about 1,000–1,500 miles. However, telegraphs also suffered degradation due to noisy channels (and noisy repeaters, some of whom were simply humans who copied a message and re-transmitted them, introducing error, or "human noise"). And, further, telegraphs, after the invention of the tube amplifier, began to use analog circuits, often combined (or "multiplexed") with telephone circuits, to carry telegraph pulses riding on an analog waveform (carrier circuits). In reality, things got even more complex than that, but the bottom line is analog circuits introduced significant limitations for telecom because of eventual degradation of the signal.

2. Unfortunately for communications between machines, there is no inherent intelligence to overcome noisy lines. Computers receive binary signals—so-called one's and zero's—and if any one of the one's are replaced by zero's because of a faulty circuit, there is trouble. The good news is that computer processes can be designed to exhibit some intelligence about the mistakes and make good statistical guesses, at some loss of "efficiency," about what the correct signal should be. This is not a trivial task, and much effort goes into designing digital communications circuits so that they are error free in practice despite a host of different obstacles that intercede. And not only are all transmission circuits noisy, but each kind of circuit—wires, coaxial cable, fiber, satellite channels, cellular radio, etc.—exhibits different noisy characteristics requiring specialized techniques to overcome.

More good news is that digital technology is not only good for data communications, but for humans as well. Once sound is digitized, as on a digital compact disk, effectively all noise can be suppressed, and the signal can be duplicated ad infinitum without significant loss. This is bad for protecting copyrights, but is good for transmitting sounds around the world and back with it sounding like it is coming from the next room—hence the clarity of phone calls from across the nation or from overseas on modern fiber-optic cables.

3. One extraordinary fact always surfaces in computer lore: the digital, programmable computer was invented by Charles Babbage and Lady Ada Lovelace in the 1830s, over one hundred years before its electronic version was implemented as a cryptomachine in World War II to decode secret German and Japanese military communications (Randell 1982). Babbage was going to build it with brass gears and steam power; he never completed it, but it was eventually built to his specifications several years ago by the British Science Museum—and it worked! Tommy Flowers, the British Post Office engineer who built that electronic version in 1943–44, based it on a prototype electronic digital telephone switch; decades later he built this switch as the British Post Office's System X. Flowers worked with Alan Turing who was responsible for the design of the Colossi—the machines that broke the Nazi cyphers in preparation for the Normandy invasion on D Day (Hinsley 1993). Turing, often cited as the father of artificial intelligence, during the war years discussed with his colleagues how his seminal and very secret code-breaking machines could evolve into a powerful device for controlling processes and manipulating information. Turing, as we noted earlier, also built the first pulse-code modulation voice telephony device in 1944 (Flowers 1976).

4. Infrastructure, originally a French word for things like railways, roads, pipelines and waterways, gained popularity in the early 1960s in the anglophone world. It has now come to mean, and is used in this text, as all elements that make industrial society work—transportation and communications systems as well as schools, farms, and factories.

5. The principle was stretched eventually to giving the railways State and Federal police powers, some still in effect today—to make arrests for trespass, to settle labor disputes, and for a period during the late nineteenth century, to maintain their own private militia.

6. Interestingly, AT&T began as an antimonopoly spin-off from Western Union's telegraph monopoly because of the corporate "leveraged buyouts" of a century ago. These same antimonopoly sentiments led to the beginnings of transport and telecommunications regulation.

7. Private toll roads were built quite extensively in the early nineteenth century after the federal government dropped financing of the first national road system. Most of these failed commercially after the steam railway was introduced. This was used to argue against the concept of public toll roads a century later (*Toll Roads and Free Roads*, U.S. Congress 1938). Those states which could justify toll freeways before the 1956 act developed infrastructure about a decade earlier than other regions, with consequent increases in land values. They were compensated for missing 90/10 financing with additional "pork barrel" projects.

8. Jay Gould, in his trade of some key railway lines to Vanderbilt for control of Western Union understood this better than most railway financiers of his day. He was quite candid about his insistence on owning telegraph contracts, pointing out that Western Union provided service free for the railways in exchange for exclusive rights of way, use of the railway employees for no charge, and free maintenance of the lines; the gross earnings of the telegraph business would not cover

its costs otherwise. Western Union got in exchange a monopoly over all telegraph communications in the railway's territory. Also, Gould was able to read the telegrams with impunity since his company had contracted to deliver the messages with a minimum of error, though with no legal obligations if there were any error as long as it could show it had performed due diligence. This Catch-22 enabled Gould to read telegrams to gain inside information for improving his trades on the stock exchanges (Thompson 1947).

9. As we noted in footnote 3, exactly one hundred years prior to the building of the Colossi, Charles Babbage and Ada Lovelace invented the "analytical engine," a device that embodied the fundamental principles of a programmable machine and used precisely the same architecture as Turing's device. Turing, von Neumann, Bush, and others apparently were aware of and influenced by these works.

10. A Pac-Man-like game was written in 1958 by John Ward for the TX0, the first transistorized computer. It was based on a robotic logic mouse maze designed by Claude Shannon.

11. In the Matters of: Amendment of Sections 64.702 of the Commission's Rules and Regulations (Third Computer Inquiry); and Policy Concerning Rates for Competitive Common Carrier Services and Facilities Authorizations Thereof; Communications Protocols under Section 64.702 of the Commission's Rules and Regulations, Report and Order, 104 FCC 2s 958 (1986).

12. Michael Dertouzos has written on the Information Marketplace; see, for example, Dertouzos 1991a and 1991b. The impact on developing countries is of interest to the authors, but is beyond the scope of this book. For one case study on telecommunications in Brazil, see Botelho et al. 1996.

13. We have relied heavily on the informed opinions and data analyses of others in preparing this chapter. In particular, we acknowledge a special debt to the work of Eric Bryjolffsson, Stephen Cohen, Charles Jonscher, Paul Krugman, Tom Malone, Laura d'Andrea Tyson, John Zysman, the National Research Council, and the MIT Commission on Industrial Productivity.

14. Similar to the academic political scientists who claimed into the 1980s that campaign contributions had no measurable impact on congressional behavior, economists challenged the notion that there was an eonomic rationale for investment in digital systems. In both cases, subsequent research has shown what was at fault was not common sense but rather the empirical measures being used to quantify these relationships.

15. By the 1990s, this issue had become a subject of political debate. The Clinton presidental campaign and presidency argued that education and training expenditures should be viewed as investments, by businesses and government

16. Alvin Toffler's promotion of the Third Wave concept has become a cultural reference point for the evolution of the information society. Toffler popularizes the findings of researchers such as Fritz Machulp, Marc Porat, and Charles Jonscher, quantifying the shift in employment over time from agriculture (first wave) to manufacturing (second wave) to service sector jobs. More recent

research has attempted to quantify the information-processing component of specific job categories, more or less successfully (third wave).

17. Networks may be defined as "a set of technical means—or infrastructures—and strategic norms—or infostructures—that actors with right of access can take the initiative to mobilize as means to set up and manage value-creating relationships." Networks are "interactions among a number of actors who *collaborate* in extracting value from their relationships, while competing in shaping the network architecture and and the specific uses being made of it" (Bressand 1992). More prosaically, the term telecommunication networks generally refers to a set of infrastructures providing either the routing or switching of telecommunications signals.

18. According to Census Bureau data, American families have maintained their income, because both spouses in more families are working at the same time. Wealthy Americans did become wealthier in the 1980s (supporting the adage "the rich get richer"), but wages for the majority stagnated in absolute terms, and declined relative to our competitors (Cohen and Zysman 1987).

19. The 180-day school year was established in the nineteenth century to enable children to help their families with summer chores and the harvest. Since fewer than three percent of Americans live on farms, the rationale for this calendar, to cite only one aspect of educational policy, is questionable.

20. Digitized images of reasonable quality consume vast quantities of bandwidth, memory, and processing. Since images (including moving images, which are even more resource-intensive) are expected to play a predominant role in future multimedia, hypermedia, and virtual reality, video-oriented computing and digital television entertainment technologies will both require orders of magnitude increases in the speed and capacity of semiconductors and other core components.

21. This section draws on material from McKnight 1993.

22. COHRS has contributed draft U.S. positions supporting open, extensible and scalable standards to national and international meetings. For example, see Gerovac et al. 1991; Schreiber 1990; Schreiber 1993; Lippman 1989; Lippman et al. 1990.

23. This is quite similar to the situation in the United States, where many federal agencies are seeking to carve out new turf or at least hold on to the turf they already control, in the National Information Infrastructure.

24. The budget for the "new social infrastructure," amounts to about 10 percent of the total. The Japanese government announced on 13 April 1993, a supplemental budget of 13 trillion yen ($131 billion) to finance the stimulus package. See Aizu 1993.

25. Koichiro Hayashi, President of NTT (Nippon Telegraph and Telephone) U.S., wrote in the *Nihon Keizai Shimbun* on 10 March 1993, emphasizing that the new U.S. administration's information policy is guided by Vice President Gore's "Information Superhighway" proposal, which Gore had pursued for a

number of years. Japan's government officials and LDP party leaders look like they are trying to "free-ride" this external factor to introduce a new policy set in Japan. It is a type of "Gaiatsu" technique often used to "change" conventional structure by introducing the perception of a foreign threat or pressure to suppress opposition.

26. This use of a foreign threat is a typical rationale for U.S., European, and Japanese technology policy initiatives—if the others are doing it, the United States (or Europe) (or Japan) must start a similar program. While this type of knee-jerk reaction may not be a sound basis for technology policy making, in fact it is reasonable to suggest that technologies identified by U.S. economic rivals as critical may also be critical to U.S. economic interests. Case-by-case examination nonetheless is required to judge whether the claim may be valid in any particular case.

27. The gain of more than 60 percent of its price within less than a month contributed significantly to the renewed dynamism of the Japanese stock market. For the Ministry of Finance, the scheduled opening sale of Japan Railways East's shares, which is currently owned by the Ministry, is the foremost concern.

28. Both the *Economist* and *Financial Times* have taken critical reviews of the European programs in the recent past, as the benefits and the costs of the programs were reassessed by the press and public in 1990. Significant financial setbacks and planned layoffs of the Dutch firm Philips and the French firm Thomson, the leading firms in the European consumer electronics market, were reported. This raises the question whether the programs are helping, or are merely an inefficient subsidy to a declining industry. In the authors' opinion, it is too soon to make that call on the economic return from the European projects of the 1980s—a clearer assessment should be possible in another five years. Competition between the Competition Directorate (DG 4) and the Information and Communication Technology Directorate (DG 13) also led to criticism of DG 13 policies.

29. The European GSM digital cellular telephone standard was developed explicitly to strengthen European industry in wireless technologies, with the intent of encouraging global adoption of the European standard. The European lead in this TDMA system is significant, but the system has some limitations and already appears technologically limited in comparison with CDMA spread spectrum personal communication networks being experimented with in the United States and elsewhere. See Laurence Hooper, "Selling Cellular. A new European system seeks to ease problems of the itinerant user," in Telecommunications, Wall Street Journal Special Report, *The Wall Street Journal*, 4 October 1991, p. R13.

30. The EUREKA-95 HDTV project was at first considered a technological success because it developed an HDTV system in record time (in less than two years prototypes were demonstrated), thwarting Japanese effort to impose an inferior first-generation HDTV system on the world.

31. These figures do not include the $57 billion in loan guarantees and $27 billion in loans made annually to small and medium-sized businesses in Japan for a

broad range of investment in commercial research, development, and manufacturing. U.S. public R&D funds are principally for defense applications and basic research, not commercial product development. Only 4 percent of Japanese public support of research is for defense technologies, while the figure for the United States is over 70 percent. The data reported is from Robert Cohen (1989) and the U.S. Department of Commerce (1990b).

32. Deregulation, privatization, and restructuring of telecommunications markets have progressed slowly. In France, the Minitel videotex system provides millions of businesses, homes, and academics with nationally standardized access to electronic information networks. Electronic yellow pages, messaging, and chat lines are popular. However, other nations have not shown interest in the Minitel, an isolated first-generation, state-subsidized information service.

33. The economic leverage gained from control of the railways was immense in the past; similar leverage is created today through control of networks. Safeguards against abuses of dominant positions, e.g., by controlling access to networks, will still be needed in the future. These issues will be touched upon in the concluding chapter.

34. This same goal is pursued by both Japanese and European government agencies (McKnight 1993).

35. The Justice Department keep fighting the original Bell patents even after the Supreme Court awarded the rights in 1888, with the case finally closed in 1915, some thirty years after it was first brought. Homer Cummings, *Federal Justice*, op. cit., ch. XV.

36. D. J. File 60-1-0. Sec. 1 as paraphrased in Cummings, op. cit., pp. 312–313. Reportedly, AT&T would literally destroy a competitor's equipment based on claims that it violated their patents.

37. Cummings, p. 342.

38. Quoted in Cummings, p. 344.

39. But not before it moved certain critical assets to AT&T, including circuits and Western Union's laboratories which became the core of Bell Labs some twelve years hence.

40. Cummings, p. 315.

41. Horwitz, op cit., p. 101.

42. Fifteen years after Staggers, railroads now carry the bulk of all freight on transcontinental routes, are reinstalling mainline track ripped up years ago, and there is serious consideration of rebuilding one of the abandoned transcontinental routes (the Milwaukee Road) because of traffic congestion.

43. See the controversy over the constitutionality of the 1990 Gun-Free School Zones Act overturned in 1995 by the Supreme Court in *U. S. v. Lopez* (--US--), *Washington Post*, April 27, 1995.

44. Brownell's diary is in the manuscript archives at Columbia University.

45. Experience with such systems in that era showed that they interconnected with the PSTN readily via automatic private branch telephone exchanges, though technically such connections were supposed to be blocked (it was almost impossible to police, so the regulatory authorities simply looked the other way).

46. Richard J. Solomon (1989b). Note on the involvement of the U.S. Government in the establishment of the Radio Corporation of America 1919 in response to foreign threats to the early U.S. electronics industry. Reprinted in several Congressional reports during 1989 and 1990.

47. Had this power been brought into question, then it would have been resolved for all time by the Supreme Court's Red Lion decision in 1966, which clearly stated that radio signals could not be regulated by the States since the frequencies could not be confined to State boundaries. How this would affect *cellular* radio with small cells is a matter for a future court decision!

48. Lee McKnight, Joseph P. Bailey, and Bruce A. Jacobson, "Modeling the Economics of Interoperability: Standards for Digital Television," *Revue de l'Economie Industrielle*, no. 75, 1 (1996), pp. 187–210.

49. "The Commission should be concerned about the availability of service to viewers rather than the profitability, or even the continued existence, of particular firms or industries" (Setzer and Levy 1991).

50. The purpose of the new committee was to recommend the design of a transmission standard for HDTV.

51. Basic sources for this discussion are from an oral history project, *The Computer Pioneers*, conducted by R. Solomon at MIT and Harvard University with John McCarthy, Brian Randell, Tommy Flowers, Gordon Brown, William H. T. Holden, and others. Toward the end of the war, the U.S. Army Security Agency (later NSA) received versions of these machines, and they were installed by Bell Labs for continued cryptographic work, probably at least for another decade. The German encryption devices they were designed to crack were used by many third world countries during the 1950s, and likely even later. For obvious reasons, the workings and details of these machines are still classified. For a not very coherent view of this early history, the basic printed sources still remain: N. Metropolis et al., *A History of Computing in the Twentieth Century* (esp. Randell's "The Colossus") (New York: Academic Press, 1980); Brian Randell, *The Origins of Digital Computers—Selected Papers* (Berlin: Springer-Verlag, 1982); T. H. Flowers, *Introduction to Exchange Systems*, Foreword and Chapter 1 (New York: Wiley, 1976).

52. Snyder, etc. Corbato on oral history tapes. Remarkably, most of its outside contracts were public, and most of the work nonclassified, funneled through the National Science Foundation and other such agencies, and via the Atomic Energy Commission, itself a secretive agency, but one which was well-known to the public. Until the records of the cold war are made fully public (on both sides of the Iron Curtain) what we know is only the tip of the iceberg, but nevertheless even $10 billion for ten years at its nascency is sufficient to understand the etiology of the paradigm shift towards computerization of industry and society.

53. We realize that the term robber baron typically connotes some dastardly intent, which is disputed by some who point to the benefits of infrastructure development and economic growth. We do not wish to debate the morality of present-day entrepreneurs and large companies intent on capturing strategic positions in the information economy, and use the term robber baron only to signify their comparable intent with the railroad, telegraph, telephone and banking capitalists of a century ago. We show in this chapter how an Open Communications Infrastructure can develop to capture the benefits while limiting the social and economic costs of development of the National Information Infrastructure.

54. George Calhoun's study of wireless access (1992) estimates that fully wireless access for 500,000 lines (at voice bandwidths) in a TDMA cellular system will require only 10 to 20 Megahertz of spectrum using currently available compression techniques. In addition, CDMA techniques could provide service without requiring additional allocation of spectrum.

References

Aaron, Henry J. 1978. *Politics and the Professors: The Great Society in Perspective.* Washington, DC: Brookings Institution.

Adelson, Andrea. 1994. "Cable Group Opposes Pactel Data Proposal." *New York Times.* February 10.

Adler, Dorothy R. 1970. *British Investment in American Railways, 1834–1898.* Charlottesville: University Press of Virginia.

Aitken, Hugh G. J. 1985. *The Continuous Wave: Technology and American Radio, 1900–1932.* Princeton: Princeton University Press.

Aizu, Izumi. 1993. "Building Japan's Information Infrastructure." *Nihon Keizai Shimbun.* April 16, 1993.

Allard, Nicholas W. 1993. "Reinventing Rate Regulation." *Federal Communication Law Journal* 46(1):63–123.

Allen, David. 1988. "New Telecommunications Services: Network Externalities and Critical Mass." *Telecommunications Policy* (September): 257–271.

Allen, Frederick Lewis. 1935. *The Lords of Creation.* Reprint Quadrangle, paper. New York: Harpers.

Alt, James E., and K. Alec Chrystal. 1983. *Political Economics.* Berkeley: University of California Press.

Altshuler, Alan. 1992. "The Politics of Deregulation," in Sapolsky, Harvey, W. Russell Neuman, Eli Noam, and Rhonda Crane, ed., *The Telecommunications Revolution: Past, Present and Future*, pp. 11–17. New York: Routledge.

Anania, Loretta. 1989. "The Politics of Integration: Telecommunications Planning in the Information Societies." Ph.D. dissertation. MIT, Cambridge, MA.

Anania, Loretta, and Richard Solomon. 1988. "Flat—the Minimalist Rate." Seventeenth Annual Telecommunications Policy Conference, Airlie, VA.

Anderson, Philip, and Michael Tushman. 1990. "Technological Discontinuities and Dominant Designs: A Cyclical Model of Technological Change." *Administrative Science Quarterly* 35:604–633.

Antonelli, Cristiano, ed. 1988. *New Information Technologies and Industrial Change.* Dordrecht: Kluwer Academic.

Antonelli, Cristiano, ed. 1992. *The Economics of Information Networks.* Amsterdam: North-Holland.

Archer, Gleason Leonard. 1938. *History of radio to 1926.* New York: American Historical Society.

Arnold, Erik, and Ken Guy. 1986. *Parallel Convergence: National Strategies in Information Technology.* Westport, CT: Quorum Books.

Aronson, Jonathan. 1992. "Telecommunications Infrastructure and U.S. International Competitiveness," in Institute for Information Studies, ed., *A National Information Network,* pp. 55–90. Queenstown, MD: Aspen Institute.

Atkin, David J., and Michael Starr. 1990. "The US Cable Communications Act Reconsidered." *Telecommunications Policy* (August): 315–323.

Axelsson, Bjoern. 1992. "Network Research—Future Issues," in Axelsson, Bjoern, and Geoffrey Easton, ed., *Industrial Networks: A New View of Reality,* pp. 237–251. London: Routledge.

Baer, Walter. 1995. "Government Investment in Telecommunications Infrastructure," in National Research Council, ed., *The Changing Nature of Telecommunications/Information Infrastructure,* pp. 179–194. Washington, DC: National Academy Press.

Bailey, J., and McKnight, Lee. 1995. "Internet Economics: What Happens When Constituencies Collide." *INET '95 Conference Proceedings.*

Banning, William P. 1945. *Commercial Broadcasting Pioneer: The WEAF Experiment: 1922–1926.* Cambridge: Harvard University Press.

Barfield, Claude E., and William A. Schambra, ed. 1986. *The Politics of Industrial Policy.* Washington, DC: American Enterprise Institute.

Barnett, Stephen R., Michael Botein, and Eli M. Noam, ed. 1988. *Law of International Telecommunications in the United States.* Nomos Verlagsgesellschaft.

Barnich, Terrence L., Craig M. Clausen, and Calvin S. Monson. 1992. "Communications Free Trade Zones: Crafting a Model for Local Exchange Competition." Illinois Commerce Commission.

Barnouw, Erik. 1966. *A Tower in Babel.* New York: Oxford University Press.

Barnouw, Erik. 1968. *The Golden Web: A History of Broadcasting in the United States.* New York: Oxford University Press.

Barnouw, Erik. 1970. *The Image Empire: A History of Broadcasting in the United States.* New York: Oxford University Press.

Baumol, William J., John C. Panzar, and Robert D. Willig. 1988. *Contestable Markets and the Theory of Industrial Structure.* New York: Harcourt, Brace, Jovanovich.

Baumol, William J., and J. Gregory Sidak. 1994. *Toward Competition in Local Telephony.* Cambridge: MIT Press.

Baumol, William J. and K. McLennan, ed. 1985. *Productivity Growth and U.S. Competitiveness.* New York: Oxford University Press.

Beebe, Lucius. 1952. *Hear the Train Blow.* New York: Dutton.

Beesley, Michael E., and Bruce Laidlaw. 1989. *The Future of Telecommunications: An Assessment of the Role of Competition in UK Policy.* London: Institute of Economic Affairs.

Bell, Daniel. 1973. *The Coming of Post-Industrial Society: A Venture in Social Forecasting.* New York: Basic Books.

Beltz, Cynthia A. 1991. *High Tech Maneuvers: The Policy Lessons of HDTV.* Washington, DC: American Enterprise Institute.

Beniger, James R. 1986a. *The Control Revolution: Technological and Economic Origins of the Information Society.* Cambridge: Harvard University Press.

Beniger, James R. 1986b. "The Information Society: Technological and Economic Origins," in Ball-Rokeach, Sandra J., and Muriel Cantor, ed., *Media, Audience and Social Structure,* pp. 51–70. Beverly Hills: Sage.

Benjamin, R. I., and M. Scott Morton. 1988. "Information Technology, Integration, and Organizational Change." *Interfaces* 18(May/June): 86–98.

Berle, Adolf. 1932. *Modern Corporations and Private Property.* New York: Commerce Clearing House.

Bernstein, Marvar. 1955. *Regulating Business by Independent Commission.* Princeton: Princeton University Press.

Berresford, John W. 1989. "The Impact of Law and Regulation on Technology: The Case History of Cellular Radio." *The Business Lawyer* 44 (May): 721–735.

Bertalanffy, Ludwig von. 1968. *General System Theory.* 1st, 3d printing. New York: Braziller.

Besen, Stanley M., Thomas G. Krattenmaker, Jr. A. Richard Metzger, and John R. Woodburg. 1984. *Misregulating Television.* Chicago: University of Chicago Press.

Besen, Stanley M., and Garth Saloner. 1989. "The Economics of Telecommunications Standards," in Crandall, Robert W., and Kenneth Flamm, ed., *Changing the Rules: Technological Change, International Competition and Regulation in Communications,* pp. 177–220. Washington, DC: Brookings Institution.

Boettinger, Henry M. 1977. *The Telephone Book: Bell, Watson, Vail and American Life, 1876–1976.* Croton-on-Hudson: Riverwood Press.

Bollier, David. 1993. "The Information Evolution: How New Information Technologies Are Spurring Complex Patterns of Change." Forum report, Program on Communications and Society, Aspen Institute.

Borrus, Michael, and François Bar. 1993. "The Future of Networking." BRIE University of California, Berkeley.

Botelho, Antonio, Robert Cohen, Lee McKnight, and Lester Thurow. 1989. "HDTV and Industrial Policy: Lessons for the 1990s." November 2. Communications Forum Seminar Notes, MIT.

Botelho, Antonio, Jose Ferro, Lee McKnight, and Antonio Manfredini. 1996. "Telecommunications in Brazil" in Eil Noam, ed. *Telecommunications in Latin America*. Oxford University Press.

Bowen, E. G. 1987. *Radar Days*. Bristol: Adam Hilger.

Bradley, Stephen P., and Jerry A. Hausman, ed. 1989. *Future Competition in Telecommunications*. Boston: Harvard Business School Press.

Branscomb, Anne W. 1983. "Communication Policy in the United States: Diversity and Pluralsim in a Competitive Marketplace," in Edgar, Patricia, and Syed A. Rahim, ed., *Communication Policy in Developed Countries*, pp. 15–56. Boston: Kegan Paul International.

Branscomb, Lewis M. 1992. "Does America Need a Technology Policy?" *Harvard Business Review* 70(2):24–33.

Braun, Mark Jerome. 1993. *AM Stereo and the FCC: Case Study of a "Marketplace" Shibboleth*. Norwood, NJ: Ablex.

Bressard, Albert, ed. 1992. *NETWORLD: The Emerging Global Networked Society*. Tokyo.

Briggs, Asa. 1982. *The Power of Steam: An Illustrated History of the World's Steam Age*. 1st. Bristol: Book Promotions.

Brock, Gerald W. 1981. *The Telecommunications Industry*. Cambridge: Harvard University Press.

Brock, Gerald W. 1994. *Telecommunications Policy for the Information Age*. Cambridge: Harvard University Press.

Bromley, D. Allan. 1990. *U.S. Technology Policy*. Washington DC: Office of Science and Technology Policy, Executive Office of the President.

Brooke, Geoffrey Mark. 1992. "The Economics of Information Technology: Explaining the Productivity Paradox." MIT.

Brooks, John. 1976. *Telephone, The First Hundred Years*. New York: Harper.

Brotman, Stuart N., ed. 1987. *The Telecommunication Deregulations Sourcebook*. Boston: Artech.

Brotman, Stuart N. 1989. "U.S. Communications Policymaking." *Telecommunications Policy* 13(4):302–308.

Brown, George E., Jr. 1993. "Technology's Dark Side." *Chronicle of Higher Education* (June 30).

Brynjolfsson, Erik. 1991a. "Information Technology and the 'Productivity Paradox': What We Know and What We Don't Know." MIT.

Brynjolfsson, Erik. 1991b. "Knowledge is Power: An Incomplete Contracts Theory of Information, Technology and Organization." MIT.

Brynjolfsson, Erik. 1994. "The Productivity Paradox." *Proceedings of the ACM*.

Brynjolfsson, Erik, and Lorin Hitt. 1993. "Is Information System Spending Productive? New Evidence and New Results." WP 3571–93. MIT.

Brynjolfsson, Erik, Thomas W. Malone, Vijay Gurbaxani, and Ajit Kambil. 1993. "Does Information Technology Lead to Smaller Firms?" MIT.

Buckingham, J. S. 1842. *The Eastern and Western States of America*. London.

Burgess, George H. 1949. *The Centennial History of the Pennsylvania Railroad Company, 1846–1946*. Philadelphia: Pennsylvania Railroad.

Bush, Vannevar. 1949. *Modern Arms and Free Men: A Discussion of the Role of Science in Preserving Democracy*. New York: Simon and Schuster.

Bush, Vannevar. 1970. *Pieces of the Action*. New York: Morrow.

Bykowsky, Mark, William Maher, and Timothy Sloan. 1991. "DRI Study: A Review," in ed., *Telecommunications in the Age of Information*, pp. C1–C25. Washington, DC: National Telecommunication and Information Administration, U.S. Department of Commerce.

Calhoun, George. 1988. *Digital Cellular Radio*. Norwood, MA: Artech.

Calhoun, George. 1992. *Wireless Access and the Local Telephone Network*. Norwood, MA: Artech.

Caporaso, James A., and David P. Levine. 1992. *Theories of Political Economy*. New York: Cambridge University Press.

Carey, John. 1985. "The Diffusion of New Telecommunications Technologies." *Telecommunications Policy* 9(2):145–158.

Carlson, Jim. 1994. "AT&T Executive Attacks Cable Industry as War over Interactive TV Heats Up." *Wall Street Journal*. January 10.

Cash, James I., Jr., and Benn R. Konsynski. 1985. "IS Redraws Competitive Boundaries." *Harvard Business Review* 85 (March–April).

Casson, Herbert N. 1910. *The History of the Telephone*. 3d. Chicago: McClurg.

Cerf, Vinton G. 1991. "Networks." *Scientific American* 265(3):72–85.

Chaffee, Steven H., and Charles R. Berger. 1987. "What Communication Scientists Do," in Berger, Charles R., and Steven H. Chaffee, ed., *Handbook of Communication Science*, pp. 99–122. Newbury Park, CA: Sage.

Chandler, Alfred D., Jr. 1969. *The Railroads*. New York: Harcourt Brace Jovanovich.

Chandler, Alfred D., Jr. 1977. *The Visible Hand*. Cambridge: Harvard University Press.

Chandler, Alfred D., Jr. 1989. *Scale and Scope: The Dynamics of Industrial Capitalism*. Cambridge: Harvard University Press.

Cherry, Colin. 1966. *On Human Communication*. Cambridge: MIT Press.

Choate, Patrick. 1990a. *Agents of Influence*. New York: Knopf.

Choate, Patrick. 1990b. "Political Advantage: Japan's Campaign for America." *Harvard Business Review* (September–October): 87–103.

Clemons, Eric K. 1991. "Evaluation of Strategic Investments in Information Technology." *Communications of the ACM* 34(1):22–36.

Clinton, William J., and Albert Gore, Jr. 1993. "Technology for America's Economic Growth: A New Diretion to Build Economic Strength." The White House.

Clockey, A. A. 1936. "Printing Telegraph Systems," in Pender, H., and K. McIwain, ed., *Electrical Engineers' Handbook*, pp. 11–25. New York: Wiley.

Coase, R. H. 1988. *The Firm, the Market, and the Law*. Chicago: University of Chicago Press.

Cochran, Thomas C. 1942. *The Age of Enterprise: A Social History of Industrial America*, revised edition 1961, orig Macmillan. New York: Harper.

Codding, George, and Anthony R. Rutkowski. 1986. *The International Telecommunication Union in a Changing World*. Norwood, MA: Artech.

Coe, Lewis. 1993. *The Telegraph: A History of Morse's Invention and Its Predecessors in the United States*. 1st. Jefferson, NC: McFarland.

Cohen, Jeffrey E. 1992. *The Politics of Telecommunications Regulation: The States and Divestiture of AT&T*. Armonk, NY: M. E. Sharpe.

Cohen, Linda. 1989. "Federal Policies for Telecommunications Research and Development," in Newberg, Paula, ed., *New Directions in Telecommunications Policy, Volume 2 Information Policy and Economic Policy*, pp. 170–186. Durham, NC: Duke University Press.

Cohen, Robert, Lee McKnight, Lester Thurow, and Antonio J. Botelho. 1989. "HDTV and Industrial Policy: Lessons for the 1990s." November 2. Seminar notes from the Communications Forum, MIT.

Cohen, Sharon. 1994. "Losing a Job to a Computer" *Associated Press*. February 23.

Cohen, Stephen S., and John Zysman. 1987. *Manufacturing Matters: The Myth of the Post-Industrial Economy*. New York: Basic.

COHRS (Committee for Open High Resolution Systems). 1990.

COHRS (Committee for Open High Resolution Systems). 1992a.

COHRS (Committee for Open High Resolution Systems). 1992b.

COHRS (Committee for Open High Resolution Systems). 1992c.

Coll, Steven. 1986. *The Deal of the Century: The Breakup of AT&T*. New York: Touchstone.

Coman, Katarine. 1930. *The Industrial History of the United States*. New York: Macmillan.

Comer, Douglas E. 1991. *Internetworking with TCP/IP*. 2d. Englewood Cliffs, NJ: Prentice-Hall.

Commission of the European Communities. 1990. "RACE '90. Research and Development in Advanced Communications Technologies in Europe." March.

Commission of the European Communities Directorate General 13. 1990a. *Perspectives for Advanced Communications in Europe: IDS 3 Business Data Communications Services*. Brussels.

Commission of the European Communities Directorate General 13. 1990b. *Perspectives for Advanced Communications in Europe: Summary Report*. Brussels.

Compaine, Benjamin. 1991. *The Future of Media Companies in the International Arena*. Cambridge: Samara Associates.

Computer Systems Policy Project. 1993. "Prospectives on the National Information Infrastructure."

Coon, Horace Campbell. 1939. *American Telephone and Telegraph: The Story of a Great Monopoly*. New York: Longmans.

Cowen, Tyler, ed. 1992. *Public Goods and Market Failures*. New Brunswick, NJ: Transaction Books.

Cowhey, Peter F. 1989. "Telecommunication and Foreign Economic Policy," in Newberg, Paula, ed., *New Directions in Telecommunications Policy, Volume 2 Information Policy and Economic Policy*, pp. 187–230. Durham, NC: Duke University Press.

Cox, Donald C. 1989. "Portable Digital Radio Communications: An Approach to Tetherless Access." (July): 30–40.

Crandall, Robert W., ed. 1991. *After the Breakup: U.S. Telecommunications in a More Competitive Era*. Washington, DC: Brookings Institution.

Crandall, Robert W. 1995. "Government Regulation and Infrastructure Development," in National Research Council, ed., *The Changing Nature of Telecommunications/Information Infrastructure*, pp. 118–124. Washington, DC: National Academy Press.

Crandall, Robert W., and Kenneth Flamm, ed. 1989. *Changing the Rules: Technological Change, International Competition and Regulation in Communications*. Washington, DC: Brookings Institution.

Crane, Rhonda. 1979. *The Politics of International Standards: France and the Color TV War*. Norwood, NJ: Ablex.

Cronin, Francis J., Elisabeth K. Colleran, Paul L. Herbert, and Steven Lewitsky. 1993. "Telecommunications and Growth: The Contribution of Telecommunications Infrastructure Investment to Aggregate and Sectoral Productivity." *Telecommunications Policy* 17(9):677–690.

Cronin, Francis J., Mark A. Gold, and Steven Lewitsky. 1992. "Telecommunications Technology, Sectoral Prices and International Competitiveness." *Telecommunications Policy* 16(7):553–575.

Cronin, Francis J., Paul Herbert, and Elisabeth Colleran. 1992. "Linking Telecommunications and Economic Competitiveness." *Telephony* (September 7, November 2).

Cronin, Francis J., Edwin B. Parker, Elisabeth K. Colleran, and Mark A. Gold. 1991. "Telecommunications Infrastructure and Economic Growth: An Analysis of Causality." *Telecommunications Policy* 15(6):529–535.

Cummings, Homer. 1937. *Federal Justice*. New York: Macmillan.

Czitrom, Daniel J. 1982. *Media and the American Mind*. Chapel Hill: University of North Carolina Press.

Daggett, Stuart. 1908. *Railroad Reorganization*. Reprint 1967. Cambridge: Harvard University Press.

Danielian, Noobar Retheos. 1939. *AT&T; The Story of Industrial Conquest*. New York: The Vanguard Press.

David, Paul A. 1990. "The Dynamo and the Computer: An Historical Perspective on the Modern Productivity Paradox." *American Economic Papers and Proceedings* (May): 355–361.

David, Paul A., and Julie Ann Bunn. 1988. "The Economics of Gateway Technologies and Network Evolution." *Information Economics and Policy* 3:165–202.

Davidson, William H. 1990a. "Analysis of Errors in Economics and Technology, Inc's the Telecommunications Infrastructure in Perspective." Submitted in response to the *NTIA Notice of Inquiry: Comprehensive Study of the Domestic Telecommunications Infrastructure*, Management Education Services Associates (MESA).

Davidson, William H. 1990b. "The Impact of New Telecommunications Policies on the Japanese Economy and Society." Submitted in response to the *NTIA Notice of Inquiry: Comprehensive Study of the Domestic Telecommunications Infrastructure*, Management Education Services Associates (MESA).

Davidson, William H., and Ronald Hubert. 1990. "A Comprehensive Assessment of the National Public Telecommunications Infrastructures." Submitted in response to the *NTIA Notice of Inquiry: Comprehensive Study of the Domestic Telecommunications Infrastructure*, Management Education Services Associates (MESA).

Davis, Clarence B., ed. 1991. *Railway Imperialism*. Westport, CT: Greenwood Press.

Davis, Lance E., and Douglass C. North. 1971. *Institutional Change and American Economic Growth*. New York: Cambridge University Press.

de Fontenay, Alain, Mark Hoffberg, Mary H. Shugard, and Robert G. White. 1987. "Local Competition and Resale of Network Services in the USA." *Telecommunications Policy* 11(1):45–57.

de Fontenay, Alain, Shugard, and Sibley, ed. 1992. *Telecommunications Demand Modeling: An Integrated View*.

De Vany, A., R. Eckert, C. Meyers, D. O'Hara, and R. Scott. 1969. "A Property System for Market Allocation of Electromagnetic Spectrum." *Stanford Law Review* (June): 1499–1561.

Defense Manufacturing Board. 1989. "Annual Report." Defense Manufacturing Board.

Degler, Carl N. 1967. *The Age of the Economic Revolution: 1876–1900*. Glenview, IL.: Scott Foresman.

DeLamarter, Richard. 1986. *Big Blue*. New York: Dodd and Mead.

Derthick, Martha, and Paul J. Quirk. 1985. *The Politics of Deregulation.* Washington, DC: Brookings Institution.

Dertouzos, Michael. 1991a. "Building the Information Marketplace." *Technology Review* (January):29–40.

Dertouzos, Michael. 1991b. "Communications, Computers and Networks." *Scientific American* 265(3):62–71.

Dertouzos, Michael L., Richard K. Lester, and Robert M. Solow. 1989. *Made in America: Regaining the Productivity Edge.* Cambridge: MIT Press.

Destler, I. M. 1992. *American Trade Politics.* Washington, DC: Institute for International Economics.

Dewing, Arthur Stone. 1926. *The Financial Policy of Corporations.* New York: Ronald Press.

Director, Mark D. 1992. *Restructuring and Expanding National Telecommunications Markets.* Washington, DC: Annenberg Washington Program.

Dizard, Wilson P., Jr. 1989. *The Coming Information Age: An Overview of Technology, Economics, and Politics.* New York: Longman.

Dordick, Herbert S. 1986. *Understanding Telecommunications.* New York: McGraw-Hill.

Dordick, Herbert S., Helen G. Bradley, and Burt Nanus. 1981. *The Emerging Network Marketplace.* Norwood, NJ: Ablex.

Dore, Ronald. 1986. *Flexible Rigidities: Industrial Policy and Structural Adjustment in the Japanese Economy 1970–80.* Stanford: Stanford University Press.

Dore, Ronald. 1987. *Taking Japan Seriously: A Confucian Perspective on Leading Economic Issues.* Stanford: Stanford University Press.

Dornbusch, Rudiger, Paul Krugman, and Yung Chul Park. 1989. *Meeting World Challenges: U.S. Manufacturing in the 1990s.* Rochester, NY: Eastman Kodak Company.

Dreher, Carl. 1977. *Sarnoff, an American Success.* New York: Quadrangle.

Drucker, Peter F. 1988. "The Coming of the New Organization." *Harvard Business Review* (Jan/Feb 1988): 45–53.

Drucker, Peter F. 1991. "The New Productivity Challenge." *Harvard Business Review* (November–December).

Duch, Raymond. 1991. *Privatizing the Economy: Telecommunications Policy in Comparative Perspective.* Ann Arbor: University of Michigan Press.

Dutton, William H. 1992. "The Ecology of Games in Telecommunications Policy," in Sapolsky, Harvey, W. Russell Neuman, Eli Noam, and Rhonda Crane, ed., *The Telecommunications Revolution: Past, Present and Future,* pp. 65–88. New York: Routledge.

Eckstein, Harry. 1975. "Case Study and Theory in Political Science," in Greenstein, Fred, and Nelson Polsby, ed., *Handbook of Political Science,* pp. 79–137. Reading, MA: Addison-Wesley.

Economedies, Nicholas. 1991. *Compatibility and Market Structure*. New York: Stern School, New York University.

Egan, Bruce L. 1991. *Information Superhighways: The Economics of Advanced Public Communications Networks*. Boston: Artech.

Egan, Bruce L., and Dennis L. Weisman. 1986. "The US Telecommunications Industry in Transition: Bypass, Regulation and the Public Interest." *Telecommunications Policy* 10(2):164–176.

Egan, Bruce L., and Steven S. Wildman. 1992. "Investing in the Telecommunications Infrastructure: Economics and Policy Considerations," in Institute for Information Studies, ed., *A National Information Network*, pp. 19–54. Queenstown, MD: Aspen Institute.

Elton, Martin C. J., ed. 1991. *Integrated Broadband Networks: The Public Policy Issues*. Amsterdam: North-Holland.

Emery, Mark J. 1991. "The Impact of Personal Communications Service on Local Exchange Carriers." Masters thesis. Massachusetts Institute of Technology.

Ergas, Henry. 1987. "Does Technology Policy Matter?" in Guile, Bruce R., and Harvey Brooks, ed., *Technology and Global Industry: Companies and Nations in the World Economy*, Washington, DC: National Academy Press.

Estrin, Deborah. 1987. "Interconnection of Private Networks: A Link Between Industrial and Telecommunications Policy." *Telecommunications Policy* 11(3): 247–258.

Fagen, M. D., ed. 1978. *History of Science and Technology in the Bell System— National Service in War and Peace (1925–1975)*. Murray Hill, NJ: Bell Telephone Laboratories.

Fallows, James M. 1989. *More Like Us: Making America Great Again*.

Fano, Robert M. 1961. *Transmission of Information: A Statistical Theory of Communications*. Cambridge: MIT Press.

Farrell, Joseph, and Garth Saloner. 1987. "Competition, Compatibility and Standards: The Economics of Horses, Penguins and Lemmings," in Gabel, H. Landis, ed., *Product Standardization and Competitive Strategy*, pp. 1–22. New York: North-Holland.

Farrell, Joseph, and Carl Shapiro. 1992. "Standard Setting in High Definition Television." University of California, Berkeley.

Faulhaber, Gerald R. 1987. *Telecommunications in Turmoil: Technology and Public Policy*. Cambridge: Ballinger.

FCC (Federal Communications Commission). 1992. "Notice of Proposed Rulemaking: Personal Communications Services." Docket No. 90–314.

FCC (Federal Communications Commission). 1979. *Report on the Policy Research Program*. Washington, DC.

FCC (Federal Communications Commission). 1988. "Tentative Decision and Further NOI in the Matter of Advanced Television and its Impact on the Existing Terrestrial Broadcast Service." September 1.

Federal Coordinator of Transportation. 1938. "Public Aids to Transportation." Vol. II. Government Printing Office.

Ferguson, Charles H. 1989. "Macroeconomic Variables, Sectoral Evidence and New Models of Industrial Performance." OECD Conference on Science, Technology and Economic Growth. Paris.

Ferguson, Charles H. 1990a. "Computers and the Coming of the U.S. Keiretsu." *Harvard Business Review* (July–August): 55–70.

Ferguson, Charles H. 1990b. "Decline or Cooperative Renewal: The Fateful Choice for American Information Technology." MIT.

Ferguson, Marjorie, ed. 1986. *New Communications Technologies and the Public Interest: Comparative Perspectives on Policy and Research.* Newbury Park, CA: Sage.

Field, Henry M. 1866. *History of the Atlantic Telegraph.* New York: Charles Scribner.

Fike, John I. 1983. *Understanding Telephone Electronics.* Carmel, CA: Sams.

Firestone, Charles. 1995. "The Search for the Holy Paradigm: Regulation the Information Infrastructure in the 21st Century," in National Research Council, ed., *The Changing Nature of Telecommunications/Information Infrastructure,* pp. 34–62. Washington, DC: National Academy Press.

Fischer, Claude S. 1991. *America Calling: A Social History of the Telephone to 1940.* Berkeley: University of California Press.

Fisher, Francis Dummer, and Arthur Melmed. 1991. "Towards a National Information Infrastructure of Education." June 10, New York University.

Flamm, Kenneth. 1987. *Targeting the Computer: Government Support and International Competition.*

Flamm, Kenneth. 1988. *Creating the Computer: Government, Industry, and High Technology.*

Flink, James J. 1970. *America Adopts the Automobile, 1895–1910.* Cambridge: MIT Press.

Flowers, Thomas H. 1976. *Introduction to Exchange Systems.* New York: Wiley.

Fogel, Robert. 1964. *Railroads and American Economic Growth.*

Fowler, Mark S., Albert Halprin, and James D. Schlichting. 1986. "Back to the Future: A Model for Telecommunications." *Federal Communications Law Journal* 38(145).

Friedman, Milton, and Ann J. Schwartz. 1963. *A Monetary History of the United States.* Princeton: Princeton University Press.

Gabel, David. 1994. "Competition in a Network Industry: The Telephone Industry, 1894–1910." *Journal of Economic History* (September).

Gabel, David, and D. Mark Kennet. 1994. "Economies of Scope in the Local Telephone Exchange Market." *Journal of Regulatory Economics* 6.

Gabel, H. Landis, ed. 1987. *Product Standardization and Competitive Strategy.* New York: North-Holland.

Galbraith, John Kenneth. 1993. *American Capitalism: The Concept of Countervailing Power.* New Brunswick, NJ: Transaction.

Garbade, Kenneth D., and William L. Silber. 1978. "Technology, Communication and the Performance of Financial Markets 1840–1975."

Gates, Paul W. 1934. *The Illinois Central and Its Colonization Work.* Harvard Economic Studies. Cambridge: Harvard University Press.

Geller, Henry. 1974. "A Modest Proposal to Reform the Federal Communications Commission." Rand Corporation.

Geller, Henry. 1989. "The Federal Structure for Telecommunciations Policy." No. 8. Benton Foundation Project on Communications & Information Policy Options, Washington Center for Policy Options.

Geller, Henry, and Donna Lampert. 1989. "Charging for Spectrum Use." No. 3. One of eight papers that comprise the Benton Foundation Project on Communications and Information Policy Options, Washington Center for Public Policy Research.

General Accounting Office. 1986. "Telephone Communications: Bypass of the Local Telephone Companies."

Gerovac, Branko, Suzanne Neil, and Lee McKnight. 1991. "Work Toward a Universal Header/Descriptor for HDTV/HRS." CCIR 11/9, September.

Gershon, Richard A. 1992. "Telephone-Cable Cross-Ownership: A Study in Policy Alternatives." *Telecommunications Policy* 16(2):110–121.

Gilder, George. 1991a. "Into the Telecosm." *Harvard Business Review* (March–April):150–161.

Gilder, George. 1991b. *Life After Television.* Knoxville, TN: Whittle Direct Books.

Gilder, George. 1992. "Cable's Secret Weapon." *Forbes*, April 18.

Gilroy, Bernard Michael. 1993. *Networking in Multinational Enterprises: The Importance of Strategic Alliances.* Critical Issues Facing the Multinational Enterprise. Columbia, SC: University of South Carolina Press.

Gleckman, Howard. 1993. "The Technology Payoff: A Sweeping Reorganization of Work Itself is Boosting Productivity." *Business Week.* June 14, pp. 57–68.

Goddard, Stephen B. 1994. *Getting There: The Epic Struggle Between Road and Rail in the American Century.* New York: Basic Books.

Goodrich, Carter. 1960. *Government Promotion of American Canals and Railroads, 1800–1890.* New York: Columbia University Press.

Gordon, John Steele. 1988. *The Scarlet Woman of Wall Street: Jay Gould, Jim Fisk, Cornelius Vanderbilt, the Erie Railway Wars and the Birth of Wall Street.* New York: Weidenfel and Nicolson.

Gordon, M. N. and R. J. Baily. 1988. *The Productivity Slowdown, Measurement Issues, and the Explosion of Computer Power*. Cambridge: National Bureau of Economic Research.

Gore, Albert. 1991. "Infrastructure in the Global Village." *Scientific American* 265(3):150–153.

Gore, Albert. 1994. "Remarks." March 21. International Telecommunication Union.

Gormley, William O. 1989. *Taming the Bureaucracy: Muscles, Prayers and Other Strategies*. Princeton: Princeton University Press.

Graham, Edward M., and Paul R. Krugman. 1991. *Foreign Direct Investment in the United States*. Washington, DC: Institute for International Economics.

Graham, Otis W., Jr. 1992. *Losing Time: The Industrial Policy Debate*. Cambridge: Harvard University Press.

Greene, Harold H. 1991. "Order: USA vs. Western Electric Co. et al." July 25. United States District Court.

Grewlich, Klaus W. 1992. *Europa Im Globalen Technologiewettlauf: Der Weltmarkt Wird Zum Binnenmarkt*. Strategien und Optionen fur die Zukunft Europas. Gutersloh, Germanry: Verlag Bertelsmann Stiftung.

Grodinsky, Julius. 1957. *Jay Gould: His Business Career 1867–1892*. Philadelphia· University of Pennsylvania Press.

Grodinsky, Julius. 1962. *Transcontinental Railway Strategy, 1869–1893: A Study of Businessmen*. Philadelphia: University of Pennsylvania Press.

Halberstam, David. 1987. *The Reckoning*. New York: Morrow.

Hall, Peter. 1986. *Governing the Economy: The Politics of State Intervention in Britain and France*. New York: Oxford University Press.

Hamdy, Walid M. 1991. "The Global Mobile Communications Market." MIT.

Hamilton, Alexander. 1964 [1792]. "Report on Manufacture," in ed., *The Reports of Alexander Hamilton*, New York: Harper and Row.

Hamsher, D. 1967. *Communication System Engineering Handbook*. New York: McGraw-Hill.

Hanson, Dirk. 1982. *The New Alchemists*. Boston: Little Brown.

Harris, L., J. Henderson, and N. Itzenberg. 1993. "Networked Health Care Delivery: Opportunities and Challenges for the 90s." October, MIT Communications Forum.

Hart, Jeffrey A. 1992. *Rival Capitalists: International Competition in the United States, Japan and Western Europe*. Ithaca: Cornell University Press.

Hart, Jeffrey A., Robert R. Reed, and François Bar. 1992. "The Building of the Internet: The Implications for the Future of Broadband Networks." Indiana University.

Hatfield, Dale N. 1995. "The Prospects for Meaningful Competition in Local Telecommunications," in National Research Council, ed., *The Changing Nature* of *Telecommunications/Information Infrastructure*, pp. 142–147. Washington, DC: National Academy Press.

Hatfield, Dale N. 1992. *The Changing Telecommunications Infrastructure and Spectrum Requirements*. Washington, DC: Annenberg Washington Program.

Haynes, R. M. 1990. "The ATM at Age Twenty: A Productivity Paradox." *National Productivity Review* 9(3):273–280.

Headrick, Daniel R. 1991. *The Invisible Weapon: Telecommunictions and International Politics, 1851–1945*. New York: Oxford University Press.

Heilbroner, Robert L. 1984. *The Economic Transformation of America*. New York: Harcourt Brace Jovanovich.

Helpman, Elhanan, and Paul R. Krugman. 1989. *Trade Policy and Market Structure*. Cambridge: MIT Press.

Hendrick, Burton J. 1937. *Bulwark of the Republic: A Biography of the Constitution*. Boston: Little Brown.

Herring, James M. 1936. *Telecommunications: Economics and Regulation*. New York: McGraw-Hill.

Hills, Jill. 1984. *Information Technology and Industrial Policy*. London: Croom Helm.

Hills, Jill. 1986. *Deregulating Telecoms: Competition and Control in the USA, Japan and Britain*. London: Pinter.

Hills, Jill. 1991. *The Democracy Gap: The Politics of Information and Communication Technologies in the United States and Europe*. New York: Greenwood Press.

Hills, Jill. 1992. "The Politics of International Telecommunications," in Sapolsky, Harvey, W. Russell Neuman, Eli Noam, and Rhonda Crane, ed., *The Telecommunications Revolution: Past, Present and Future*, pp. 120–139. New York: Routledge.

Hilton, George, and John Due. 1960. *The Electric Interurban Railways in America*. Stanford: Stanford University Press.

Hinsley, F. H. 1993. *Code Breakers: The Inside Story of Bletchley Park*. New York: Oxford University Press.

Hirschman, Albert O. 1991. *The Rhetoric of Reaction*. Cambridge: Harvard University Press.

Hofstadter, Douglas R. 1979. *Gödel, Escher, Bach*. New York: Basic Books.

Holbrook, Stewart H. 1947. *The Story of American Railroads*. New York: Crown.

Horwitz, Robert Brett. 1989. *The Irony of Regulatory Reform*. New York: Oxford University Press.

Housel, Thomas, and William H. Davidson. 1990. "The Development of Information Services in France: Proactive Public Policies and Their Social Impact." Submitted in response to the *NTIA Notice of Inquiry: Comprehensive Study of the Domestic Telecommunications Infrastructure*, Management Education Services Associates (MESA).

Huber, Peter W. 1987. *The Geodesic Network*. Washington, DC: United States Department of Justice.

Huber, Peter W., Michael K. Kellogg, and John Thorne. 1993. *The Geodesic Network II: 1993 Report on Competition in the Telephone Industry*. Washington, DC: The Geodesic Company.

Hufbauer, Gary Clyde, ed. 1989. *The Free Trade Debate*. New York: Priority Press.

Illinois Commerce Commision. 1992. "Local Competition and Interconnection." Staff report. Illinois Commerce Commission.

Imai, Ken-ichi. 1989. "Japan's Coporate Networks," in Rosovsky, Henry, ed., *The Political Economy of Japan*. Palo Alto: Stanford University Press.

Inouye, Daniel K. 1991. *Information Services Diversity Act of 1991*. Washington, DC: United States Senate.

Institute for Information Studies. 1992. *A National Information Network*. Queenstown, MD: Aspen Institute.

International Telecommunication Union. 1986. *Information, Telecommunications and Development*. Geneva.

Jackman, William T. 1916. *The Development of Transportation in Modern England*. London: Case.

Jackson, Charles Lee. 1977. "Technology for Spectrum Markets." MIT.

Jacobson, Bruce A. 1993. "The Economics of Interoperability: Modeling High Resolution Imaging in the Computer, Consumer Electronics, Broadcast and Cable Television Industries." Masters Thesis. MIT.

Jensen, Oliver. 1975. *Railroads in America*. New York: American Heritage.

Johnson, Chalmers. 1982. *MITI and the Japanese Miracle: The Growth of Industrial Policy, 1925–1975*. Stanford: Stanford University Press.

Johnson, Chalmers, ed. 1984. *The Industrial Policy Debate*. San Francisco: ICS Press.

Johnson, Chalmers. 1995. *Japan: Who Governs? The Rise of the Developmental State*. New York: Norton.

Johnson, Emory R. 1926. *Principles of Railroad Transportation*. 6th. ed. New York: D. Appleton.

Johnson, Leland L., and David P. Reed. 1992. "Telephone Company Entry into Cable Television." *Telecommunications Policy* 16(2):122–134.

Jones, Peter d'A. 1968. *The Robber Barons Revisited*. Boston: D. C. Heath.

Jonscher, Charles. 1983. "Information Resources and Economic Productivity." *Information Economics and Policy* 1(1):13–35.

Jonscher, Charles. 1986. "Information Technology and the United States Economy," in Faulhaber, Gerald, Eli Noam, and Roberta Tasley, ed., *Services in*

Transition: The Impact of Information Technologies on Services, Cambridge, MA: Ballinger.

Josephson, Matthew. 1934. *The Robber Barons: The Great American Capitalists, 1861–1901*. New York: Harcourt, Brace.

Joskow, Paul L., and Roger C. Noll. 1981. "Regulation in Theory and Practice: An Overview," in Fromm, Gary, ed., *Studies in Public Regulation*, pp. 1–65. Cambridge: MIT Press.

Judice, C. N., E. J. Addeo, M. I. Eiger, and H. L. Lemberg. 1986. *Video on Demand: A Wideband Service or Myth?* Morristown, NJ: Bell Communications Research.

Kahin, Brian, ed. 1992. *Building Information Infrastructure*. New York: McGraw-Hill.

Kahin, Brian, ed. 1994. *Information Infrastructure: A Sourcebook*. Cambridge, MA: Harvard University John F. Kennedy School of Government Center for Science and International Affairs.

Kahn, Alfred E. 1970. *The Economics of Regulation, Volume 1: Economic Principles*. New York: Wiley.

Kahn, Alfred E. 1971. *The Economics of Regulation, Volume 2: Institutional Issues*. New York: Wiley.

Kahn, Robert E. 1995. "Economic Dividents of Government Investment in Research and Technology Development," in National Research Council, ed., *The Changing Nature of Telecommunications/Information Infrastructure*, pp. 207–214. Washington, DC: National Academy Press.

Kapor, Mitchell. 1991. "Building the Open Road: Policies for the National Public Network." Electronic Frontier Foundation.

Keen, Peter G. 1988. *Competing in Time: Using Telecommunications for Competitive Advantage*. Cambridge: Ballinger.

Kellogg, Michael K., John Thorne, and Peter Huber. 1992. *Federal Telecommunications Law*. Boston: Little Brown.

Kidwell, Peggy Aldrich. 1994. *Landmarks in Digital Computing*. 1st. Washington: Smithsonian Institution Press.

Kirkland, Edward Chase. 1965. *Charles Francis Adams, Jr.: The Patrician at Bay*. Cambridge: Harvard University Press.

Klebnikov, Paul. 1993. "Le folies HDTV." *Forbes*. July 19), pp. 65–69.

Klein, Maury. 1987. *Union Pacific: the Rebirth 1894–1969*. New York: Doubleday.

Klopfenstein, Bruce C., and David Sedman. 1990. "Technical Standards and the Marketplace: The Case of AM Stereo." *Journal of Broadcasting and Electronic Media* 34(2):171–194.

Kolko, Gabriel. 1963. *The Triumph of Conservatism: A Reinterpretation of American History, 1900–1916*. Glencoe: Free Press.

Kolko, Gabriel. 1965. *Railroads and Regulation: 1877–1916*. Princeton: Princeton University Press.

Kraemer, Joseph S. 1995. "The Realities of Convergence." EDS Management Consulting Services.

Krattenmaker, Thomas G. 1989. "Compatibility Standards and Telecommunications Policy," in Newberg, Paula, ed., *New Directions in Telecommunications Policy, Volume 2 Information Policy and Economic Policy*, pp. 290–295. Durham, NC: Duke University Press.

Krugman, Paul R. 1983. "Targeted Industrial Policies," in ed., *Industrial Change and Public Policy*, Kansas City: Federal Reserve Bank.

Krugman, Paul R. 1987. "Is Free Trade Passé?" *Journal of Economic Perspectives* 1(2):131–144.

Krugman, Paul R. 1990. *Rethinking International Trade*. Cambridge: MIT Press.

Krugman, Paul R. 1994a. *The Age of Diminished Expectations*. Cambridge: MIT Press.

Krugman, Paul R. 1994b. *Peddling Prosperity: Economic Sense and Nonsense in The Age of Diminished Expectations*. New York: Norton.

Kurisaki, Yoshiko. 1993. "Globalization or Regionalization? An Observation of Current PTO Activities." *Telecommunications Policy* 17(9):699–706.

Kuttner, Robert. 1991. *The End of Laissez-Faire*. New York: Knopf.

Kuttner, Robert. 1992. "Facing Up to Industrial Policy." *New York Times Sunday Magazine*. April 10, pp. 22–42.

Kuznets, Simon. 1971. *Economic Growth of Nations: Total Output and Production Structure*. Cambridge: Harvard University Press.

Landauer, Thomas K. 1995. *The Trouble with Computers: Usefulness, Usability, and Productivity*. Cambridge: MIT Press.

Landes, David S. 1969. *The Unbound Prometheus: Technological Change and Industrial Development in Western Europe from 1750 to the Present*. New York: Cambridge University Press.

Lardner, James. 1987. *Fast Forward: Hollywood, the Japanese, and the Onslaught of the VCR*. New York: Norton.

Latham, Earl. 1959. *The Politics of Railroad Coordination, 1933–1936*. Cambridge: Harvard University Press.

Lazonick, William. 1991. *Business Organization and the Myth of the Market Economy*. New York: Cambridge University Press.

Lebow, Irwin. 1995. *Information Highways and Byways: From the Telegraph to the Twenty-First Century*. New York: IEEE Press.

Lee, Alfred. 1988. "Land Mobile Radio Services," in Shapleigh, John C., ed., *Telecomm 2000: Charting the Course for a New Century*, pp. 283–304. Washington, DC: National Telecommunication and Information Administration, U.S. Department of Commerce.

Leebaert, Derek, ed. 1991. *Technology 2001: The Future of Commuting and Communications*. Cambridge: MIT Press.

Lensen, Anton. 1992. *Concentration in the Media Industry: The European Community and Mass Media Regulation*. Washington, DC: Annenberg Washington Program.

Levy, Jonah D., and Richard J. Samuels. 1989. "Institutions and Innovation: Research Collaboration as Technology Strategy in Japan." MIT.

Lewis, Tom. 1991. *Empire of the Air, The Men Who Made Radio*. New York: Harper.

Libicki, Martin C. 1993. "The Common Byte: Why Execellent Technology Standards are Absolutely Essential and Utterly Impossible." Harvard University Program on Information Resources Policy.

Lindley, Lester G. 1971. *The Constitution Faces Technology: The Relationship of the National Government to the Telegraph*.

Lippman, Andrew. 1986. *Fifth Generation Television*. Cambridge: Media Laboratory.

Lippman, Andrew. 1989. "Forget Television Sets: An Action Memo for Representative Markey." MIT Media Lab.

Locklin, D. Philip. 1954. *Economics of Transportation*. 4th. Homewood: Richard D. Irwin.

Loveman, G. W. 1990. "An Assessment of the Productivity Impact of Information Technologies." MIT Sloan School of Management.

Lowi, Theodore J. 1969. *The End of Liberalism: Ideology, Policy and the Crisis of Public Authority*. New York: W. W. Norton.

Lucky, Robert. 1995. "The Evolution of Telecommunications Infrastructure," in National Research Council, ed., *The Changing Nature of Telecommunications/ Information Infrastructure*, pp. 25–33. Washington, DC: National Academy Press.

MacAvoy, Paul W. 1965. *The Economic Effects of Regulation: The Trunk-Line Railroad Cartels and the Interstate Commerce Commission before 1900*. Cambridge: MIT Press.

MacAvoy, Paul W. 1992. *Industry Regulation and the Performance of the American Economy*. New York: W. W. Norton.

Machlup, Fritz. 1962. *The Production and Distribution of Knowledge in the United States*. Princeton: Princeton University Press.

Machlup, Fritz. 1980, 1982, 1984. *Knowledge: Its Creation, Distribution and Economic Significance, Volumes I–III*. Princeton: Princeton University Press.

Malone, Thomas W., and John F. Rockart. 1991. "Computers, Networks and the Corporation." *Scientific American* 265(3):128–137.

Malone, Thomas W., Joann Yates, and Richard I. Benjamin. 1987. "Electronic Markets and Electronic Hierarchies: Effects of New Information Technologies on

Market Structures and Corporate Strategies." *Communications of the ACM* 30 (June): 484–497.

Malone, Thomas W., Joann Yates, and Richard I. Benjamin. 1988. "The Logic of Electronic Markets." *Harvard Business Review* 67 (May/June): 166–172.

Mansell, Robin. 1993. *The New Telecommunications: A Political Economy of Network Evolution.* Newbury Park, CA: Sage.

Markoff, John. 1993. "Building the Electronic Superhighway." *New York Times.* January 24.

Martin, Albro. 1971. *Enterprise Denied.* New York: Columbia University Press.

Martin, Stephen. 1994. *Industrial Economics: Economic Analysis and Public Policy.* New York: Macmillan.

Mathison, Stuart L. 1970. *Computers and Telecommunications: Issues in Public Policy.* Englewood Cliffs, NJ: Prentice-Hall.

Mazlish, Bruce, ed. 1965. *The Railroad and the Space Program.* Cambridge: MIT Press.

McAdams, Alan K. 1989. "U.S. International Competitiveness: Engineering a Systems Solution." Cornell University.

McCraw, Thomas K. 1984. *Prophets of Regulation.* Cambridge: Harvard University Press.

McCraw, Thomas K. 1986. "Mercantilism and the Market: Antecedents of American Industrial Policy," in Barfield, Claude E., and William A. Schambra, ed., *The Politics of Industrial Policy,* pp. 33–62. Washington, DC: American Enterprise Institute.

McGarity, Thomas O. 1991. *Reinventing Rationality: The Role of Regulatory Analysis in the Federal Bureaucracy.* New York: Cambridge University Press.

McGarty, Terrence P. 1993a. "Access Policy and the Changing Telecommunications Infrastructures." September. Telecommunications Policy Research Conference.

McGarty, Terrence P. 1993b. "Access to the Local Loop." March. John F. Kennedy School of Government, Harvard University.

McGarty, Terrence P., and Sara J. McGarty. 1991. "Information Architectures and Infrastructures: Value Creation and Transfer." September. Telecommunications Policy Research Conference.

McHugh, F. Joseph. 1983. "Congressional Attempts to Revise the Communications Act of 1934," in Rubin, Michael Rogers, ed., *Information Economics and Policy in the United States,* pp. 161–173. Littleton, CO: Libraries Unlimited.

McKenna, Richard. 1985. "Preemption Under the Communications Act." *Federal Communications Law Journal* 37(1):1–69.

McKnight, Lee. 1989a. "Ad Hoc Corporatism and Expert Commissions." American Political Science Association Annual Meeting. Washington, DC.

McKnight, Lee. 1989b. "HDTV and the Technopolitics of Standardization." *Project PROMETHEE Perspectives* (June): 15–20.

McKnight, Lee. 1989c. "The Making of Industrial Policy." Doctoral dissertation, MIT.

McKnight, Lee. 1990. "Global Internetworks." Information Infrastructure for the 1990s Workshop, Harvard University.

McKnight, Lee. 1992. "European and Japanese Research Networks: Cooperating to Compete," in Kahin, Brian, ed., *Building Information Infrastructure*, pp. 46–58. New York: McGraw-Hill.

McKnight, Lee. 1993. "European and Japanese Research Networks: Cooperating to Compete," in Kahin, Brian, ed. *Building Information Infrastructure*. New York: McGraw-Hill.

McKnight, Lee, and W. Russell Neuman. 1995. "Technology Policy and National Information Infrastructure," in Drake, William, ed., *The New Information Infrastructure: Strategies for U.S. Policy*, New York: Twentieth Century Fund Press.

McKnight, Lee, Joseph P. Bailey, and Bruce A. Jacobson. 1996. "Modeling the Economics of Interoperability: Standards for Digital Television," *Revue de l'Economie Industrielle*, no. 75, 1, pp. 187–210.

Melody, William H. 1991. *Manufacturing in the Global Information Economy*. CIRCIT Newsletter. Melbourne, Australia.

Mercer, Lloyd J. 1982. *Railroads and Land Grant Policy: A Study in Government Intervention*. New York: Academic Press.

Mestmaecker, Ernst-Joachim, ed. 1987. *Transborder Telecommunications Law and Economics*. Baden-Baden: Nomos.

Mines, Christopher W. 1993. "Transnational Investments in Mobile Telephone Systems: Toward Global Telephone Companies." Harvard University Program on Information Resources Policy.

Moir, Brian R. 1990. *The Status of the US Telecommunications Infrastructure and the Telecommunications Policy Act of 1990*. Washington, DC: International Communications Association.

Montgomery, William Page, Lee L. Selwyn, and Paul S. Keller. 1990. "The Telecommunications Infrastrucutre in Perspective." Prepared for the Consumer Federation of America and the International Communications Association, Economics and Technology, Inc.

Morita, Akio. 1986. *Made in Japan*. New York: Signet.

Morita, Akio, and Shintaro Ishihara. 1990. *The Japan that Can Say "No"*.

Morris, Charles R., and Charles H. Ferguson. 1993. "How Architecture Wins Technology Wars." *Harvard Business Review* (March–April): 86–96.

Morrison, E. R., and C. J. Berndt. 1992. "High-Tech Capital Formation and Economic Performance in U.S. Manufacturing Industries: An Exploratory Analysis." No. 3419, April. Working Paper, MIT Sloan School of Management.

Mowery, David C. 1989. *Technology and the Pursuit of Economic Growth.* New York: Cambridge University Press.

Mueller, Milton. 1991. "Telecommunications Development Models and Telephone History." *ICTM News* 2(2):8–10.

Mueller, Milton. 1993a. "New Zealand's Revolution in Spectrum Management." *Information Economics and Policy* 5(2):159–177.

Mueller, Milton. 1993b. *Telephone Companies in Paradise: A Case Study in Telecommunication Regulation.* New Brunswick, NJ: Transaction.

Mulgan, Geoff J. 1991. *Communication and Control: Networks and the New Economies of Communication.* New York: Guilford.

Muller, Joachim W. 1990. *European Collaboration in Advanced Technology.* New York: Elsevier.

Nasar, Sylvia. 1992. "Why International Statistical Comparisons Don't Work." *New York Times.* March 8.

Nathan, Richard P. 1988. *Social Science in Government.* New York: Basic Books.

National Academy of Engineering. 1988. "The Technological Dimensions of International Competitiveness."

National Research Council. 1990. *Crossroads of Information Technology Standards.* Washington, DC: National Academy Press.

National Research Council. 1991. *Keeping the U.S. Computer Industry Competitive: Systems Integration.* Washington, DC: National Academy Press.

National Research Council, ed. 1994. *Organizational Linkages: Understanding the Productivity Paradox.* Washington, DC: National Academy Press.

National Research Council. 1994. *Realizing the Future: The Internet and Beyond.* Washington, DC: National Academy Press.

National Research Council. 1995. *The Changing Nature of Telecommunications/Information Infrastructure.* Washington, DC: National Academy Press.

National Telecommunications and Information Administration. 1987. *NTIA Trade Report: Assessing the Effects of Changing the AT&T Antitrust Consent Decree.* Washington, DC.

National Telecommunications and Information Administration. 1988. *Telecom 2000: Charting the Course for a New Century.* Washington, DC.

National Telecommunications and Information Administration. 1990. *Comprehensive Study of Domestic Telecommunications Infrastructure.* Washington, DC.

National Telecommunications and Information Administration. 1991a. *Telecommunications in the Age of Information.* Washington, DC.

National Telecommunications and Information Administration. 1991b. *U.S. Spectrum Policy: Agenda for the Future.* Washington, DC.

National Telecommunications and Information Administration. 1993. *Globalization of the Mass Media.* Washington, DC.

Nau, Henry R. 1990. *The Myth of America's Decline: Leading the World Economy into the 1990s*. New York: Oxford University Press.

Negroponte, Nicholas P. 1991. "Products and Services for Computer Networks." *Scientific American* 265(3):106–115.

Negroponte, Nicholas P. 1995. *Being Digital*. New York: Knopf.

Neil, Suzanne, Lee McKnight, and Joseph Bailey. 1995. "The Government's Role in the HDTV Standards Process: Model or Aberration?" in Kahin, Brian, ed., *Standards, Policy and Information Infrastructure*. Cambridge: MIT Press.

Nelson, Richard. 1990. "U.S. Technological Leadership: Where It Come From and Where Did it Go?" *Research Policy* 19:117–1132.

Neuman, W. Russell. 1988. "Hearings on New Directions of Video Technology." June 23. Testimony before the Subcommittee on Science, Space and Technology, U.S. House of Representatives, Advanced Television Research Project, MIT Media Laboratory.

Neuman, W. Russell. 1991. *The Future of the Mass Audience*. New York: Cambridge University Press.

Neuman, W. Russell. 1992. "Communications Policy in Crisis," in Sapolsky, Harvey, W. Russell Neuman, Eli Noam, and Rhonda Crane, ed., *The Telecommunications Revolution: Past, Present and Future*, pp. 199–205. New York: Routledge.

Neuman, W. Russell, Lee McKnight, and Richard Jay Solomon. 1993. "The Politics of a Paradigm Shift: Telecommunications Regulation and the Communications Revolution." *Political Communication* 10:77–94.

Neustadt, Richard E., and Ernest R. May. 1986. *Thinking in Time: The Uses of Hisory for Decision Makers*. New York: Free Press.

Nevins, Allan, and Henry Steele Commager. 1945. *A Short History of the United States*. New York: Random House.

New York Times. 1992. "Industrial Policy as Sloppy Slogan." *New York Times*. February 12.

Nightingale, John. 1988. "Information and Productivity: An Attempted Replication of an Empirical Exercise." *Information Economics and Policy* 3:55–67.

Noam, Eli. 1989. "Network Pluralism and Regulatory Pluralism," in Newberg, Paula, ed., *New Directions in Telecommunications Policy, Volume 1 Regulatory Policy : Telephony and Mass Media*, pp. 66–91. Durham, NC: Duke University Press.

Noam, Eli. 1992. "Beyond the Golden Age of the Public Network," in Sapolsky, Harvey, W. Russell Neuman, Eli Noam, and Rhonda Crane, ed., *The Telecommunications Revolution: Past, Present and Future*, pp. 6–10. New York: Routledge.

Noam, Eli. 1994a. "Beyond Liberalization: From the Network of Networks to the System of Systems." *Telecommunications Policy* 18(4):286–294.

Noam, Eli. 1994b. "Beyond Liberalization II: The Impending Doom of Common Carriage." *Telecommunications Policy* 18(6):435–452.

Noam, Eli. 1995a. "Beyond Liberalization III: Reforming Universal Service." *Telecommunications Policy* 19(1).

Noam, Eli. 1995b. "Economic Ramifications of the Need for Universal Telecommunications Service," in National Research Council, ed., *The Changing Nature of Telecommunications/Information Infrastructure*, pp. 161–164. Washington, DC: National Academy Press.

Noam, Eli, and Gerard Pogorel, ed. 1992. *Asymmetric Deregulation: The Dynamics of Telecommunications Policies in Europe and the United States*. Norwood, NJ: Ablex.

Noble, David F. 1979. *America by Design: Science, Technology and the Rise of Corporate Capitalism*. New York: Oxford University Press.

Noll, Roger. 1985a. "Government Regulatory Behavior: A Multidisciplinary Survey and Synthesis," in Noll, Roger, ed., *Regulatory Policy and the Social Sciences*, pp. 9–63. Berkeley: University of California Press.

Noll, Roger, ed. 1985b. *Regulatory Policy and the Social Sciences*. Berkeley: University of California Press.

Noll, Roger. 1989. "Telecommunication Regulation in the 1990s," in Newberg, Paula, ed., *New Directions in Telecommunications Policy, Volume 1 Regulatory Policy: Telephony and Mass Media*, pp. 11–48. Durham, NC: Duke University Press.

Noll, Roger G., and Bruce M. Owen, ed. 1983. *The Political Economy of Deregulation: Interest Groups in the Regulatory Process*. Washington, DC: American Enterprise Institute.

Nonaka, Ikujiro. 1995. *The Knowledge-Creating Company: How Japanese Companies Create the Dynamics of Innovation*. New York: Oxford University Press.

Nora, Simon, and Alain Minc. 1980. *The Computerization of Society*. Cambridge: MIT Press.

North, Douglass C. 1981. *Structure and Change in Economic History*. New York: W. W. Norton.

North, Douglass C. 1990. *Institutions, Institutional Change and Economic Performance*. New York: Cambridge University Press.

North, Douglass C. 1993. "Toward a Theory of Institutional Change," in Barnett, William A., Melvin J. Hinich, and Norman J. Schofield, ed., *Political Economy: Institutions, Competition and Representation*, pp. 61–69. New York: Cambridge University Press.

Nye, Joseph S. 1990. *Bound to Lead: The Changing Nature of American Power*. New York: Basic Books.

O'Brien, Patrick. 1977. *The New Economic History of the Railways*. New York: St. Martin's.

O'Connor, Richard. 1962. *Gould's Millions*. Garden City, NJ: Doubleday.

Oettinger, Anthony. 1977. *High and Low Politics: Information Resources for the 80s*. Cambridge: Ballinger.

Office of Technology Assessment, ed. 1990. *Critical Connections: Communication for the Future*. Washington, DC.

Office of Technology Assessment. 1994. "Electronic Enterprises: Looking to the Future." OTA-TCT-600. Washington, DC.

Ogan, Christine L. 1992. "Communications Policy Options in an Era of Rapid Technological Change." *Telecommunications Policy* (September): 565–575.

Okimoto, Daniel I. 1989. *Between MITI and the Market: Japanese Industrial Policy for High Technology*. Stanford: Stanford University Press.

Olson, Mancur. 1982. *The Rise and Decline of Nations: Economic Growth, Stagflation, and Social Rigidities*. New Haven: Yale University Press.

Organization for Economic Cooperation and Development (OECD). 1990. *Trade in Information, Computer and Communications Services*. Directorate on Information, Computer, Communications Policy. Paris.

Organization for Economic Cooperation and Development (OECD). 1991. *Telecommunications Equipment: Changing Markets and Trade Structures*. Directorate on Information, Computer, Communications Policy. Paris.

Oslin, George P. 1992. *The Story of Telecommunications*. Macon: Mercer University Press.

Ostrom, Elinor. 1990. *Governing the Commons: The Evolution of Institutions for Collective Action*. New York: Cambridge University Press.

OTA. See Office of Technology Assessment.

Owen, Bruce M., and Ronald Braeutigam. 1978. *The Regulation Game: Strategic Use of the Administrative Process*. Cambridge: Ballinger.

Owen, Bruce M., and Steven S. Wildman. 1992. *Video Economics*. Cambridge: Harvard University Press.

Oxley, Alan. 1991. "International Trade in Telecommunications Services: The Pressure of Free Trade Paradigms." No. 12. Center for International Research on Communication and Information Technologies.

Paglin, Max. 1989. *A Legislative History of the Communications Act of 1934*. New York: Oxford University Press.

Paine, Albert Bigelow. 1921. *Theodore N. Vail: In One Man's Life*. New York: Harper and Row.

Pepper, Robert M. 1988. "Through the Looking Glass: Integrated Broadband Networks, Regulatory Policy and Institutional Change." November. OPP Working Paper Series no. 24, Federal Communications Commission Office of Plans and Policy.

Phillips, Almarin. 1991. "Changing Markets and Institutional Inertia: A Review of US Telecommunications Policy." *Telecommunications Policy* 15(1):49–61.

Phillips, Kevin P. 1992. "U.S. Industrial Policy: Inevitable and Ineffective." *Harvard Business Review* 70(4):104–112.

Pier, Arthur S. 1953. *Forbes: Telephone Pioneer*. New York: Dodd Mead.

Pierce, John R. 1990. *Signals: The Science of Telecommunications*. New York: Scientific American Library.

Piore, Michael J., and Charles F. Sabel. 1984. *The Second Industrial Divide*. New York: Basic Books.

Pool, Ithiel de Sola. 1977. "Foresight and Hindsight: The Case of the Telephone," in Pool, Ithiel de Sola, ed. *The Social Impact of the Telephone*, Cambridge: MIT Press.

Pool, Ithiel de Sola. 1983. *Technologies of Freedom*. Cambridge: Harvard University Press.

Pool, Ithiel de Sola, Hiroshi Inose, Nozomu Takasaki, and Roger Hurwitz. 1984. *Communications Flows: A Census in the United States and Japan*. Amsterdam: Elsevier North-Holland.

Pool, Ithiel de Sola, and Richard J. Solomon. 1980. "Intellectual Property and Transborder Data Flows." *Stanford Journal of International Law* .

Porat, Marc U. 1977. *The Information Economy*. Washington, DC: Government Printing Office.

Porter, Michael E. 1980. *Competitive Strategy: Techniques for Analyzing Industries and Competitors*. New York: Free Press.

Porter, Michael E. 1990. *The Competitive Advantage of Nations*. New York: Free Press.

Porter, Michael E. 1992. "On Thinking About Deregulaton and Competition," in Sapolsky, Harvey, W. Russell Neuman, Eli Noam, and Rhonda Crane, ed., *The Telecommunications Revolution: Past, Present and Future*, pp. 39–44. New York: Routledge.

Prestowitz, Clyde V., Jr. 1988. *Trading Places: How We Allowed Japan to Take the Lead*. New York: Basic Books.

Prestowitz, Clyde V., Jr. 1991. "More Trade is Better than Free Trade." *Technology Review* (April): 23–29.

Quiett, Glenn Chesney. 1934. *They Built the West: An Epic of Rails and Cities*. New York: D. Appleton-Century.

Rae, John Bell. 1979. *The Development of Railway Land Subsidy Policy in the United States*. Reprint of author's thesis, Brown University, 1936. New York: Arno Press.

Randell, Brian, ed. 1982. *The Origins of Digital Computers: Selected Papers*. 3d. Edition. Berlin: Springer-Verlag.

Reed, David P. 1992a. "Putting It All Together: The Cost Structure of Personal Communication Services." November. OPP Working Paper Series no. 28, Federal Communications Commission Office of Plans and Policy.

Reed, David P. 1992b. *Residential Fiber Optic Networks: An Engineering and Economic Analysis.* Norwood, MA: Artech.

Reed, M. C., ed. 1969. *Railways in the Victorian Economy.* Newton Abbot: David and Charles.

Reich, Robert B. 1989. "The Quiet Path to Technological Preeminence." *Scientific American* 261(4):41–47.

Reich, Robert B. 1990. "Who Is Us?" *Harvard Business Review* (January–February): 53–64.

Reich, Robert B. 1991a. "Who Is Them?" *Harvard Business Review* (March–April): 77–89.

Reich, Robert B. 1991b. *The Work of Nations: Preparing Ourselves for 21st-Century Capitalism.* New York: Knopf.

Rhoads, Charles Stanley. 1924. *Telegraphy and Telephony with Railroad Applications.* Railwaymen's Handbook Series. New York: Simmons-Boardman.

Rhodes, Frederick Leland. 1929. *Beginnings of Telephony.* New York: Harper.

Rice, John F., ed. 1990. *HDTV: The Politics, Policies and Economics of Tomorrow's TV.* New York: Union Square Press.

Riegel, Robert Edgar. 1926. *The Story of the Western Railroads from 1852 Through the Reign of the Giants.* New York: Macmillan.

Roach, Stephen S. 1991. "Services Under Seige." *Harvard Business Review* (September October): 82–91.

Roach, Stephen S. 1992. *Assessing the Impacts of A Productivity-Led Recovery.* New York: Morgan Stanley.

Roberts, L. G. 1978. "The Evolution of Packet Switching." Proceedings of the IEEE (November): 1397–1407.

Robinson, Glen O. 1989. "The Federal Communications Act: An Essay on Origins and Regulatory Purpose," in Paglin, Max, ed., *A Legislative History of the Communications Act of 1934*, pp. 3–24. New York: Oxford University Press.

Rogers, Everett M. 1983. *Diffusion of Innovations.* New York: Free Press.

Rohlfs, Jeffrey H., Charles L. Jackson, and Tracey E. Kelly. 1991. "Estimate of the Loss to the United States Caused by the FCC's Delay in Licensing Cellular Telecommunications." National Economic Research Associates.

Rosenberg, Nathan, and L. E. Birdzell Jr. 1986. *How the West Grew Rich: The Economic Transformation of the Industrial Revolution.* New York: Basic Books.

Rosenberg, Nathan, Ralph Landau, and David C. Mowery, ed. 1992. *Technology and the Wealth of Nations.* Stanford: Stanford University Press.

Rothstein, Russell, and Lee McKnight. 1995. "Connecting Schools to the NII." *T.H.E. Journal.* In Press.

Salamon, L. M., and J. S. Siegried. 1977. "Economic Power and Political Influence: The Impact of Industry Structure on Public Policy." *American Political Science Review* 71:1026–1043.

Sapolsky, Harvey, W. Russell Neuman, Eli Noam, and Rhonda Crane. 1992. *The Telecommunications Revolution: Past, Present and Future.* New York: Routledge.

Scherer, F. M. 1992. *International High Tech Competition.* Cambridge: Harvard University Press.

Schilling, Donald L. et al. 1991. "Broadband CDMA for Personal Communications Systems." *IEEE Communications* 29(11):86–93.

Schilling, Donald L., Laurence B. Milstein, Raymond L. Pickholtz, Marvin Kullback, and Frank Miller. 1991. "Spread Spectrum for Commercial Communications." *IEEE Communications* (4):66–79.

Schivelbusch, Wolfgang. 1986. *The Railway Journey: The Industrialization of Time and Space in the 19th Century.* Berkeley: University of California Press.

Schlesinger, Arthur M., Jr. 1986. *The Cycles of American History.* Boston: Houghton Mifflin.

Schlossstein, Steven. 1990. *The End of the American Century.* New York: Congdon and Weed.

Schmitz, Hubert, and Jose Cassiolato. 1992a. "Fostering Hi-Tech Industries in Developing Countries," in Schmitz, Hubert, and Jose Cassiolato, ed., *Hi-Tech for Industrial Development: Lessons from the Brazilian Experience in Electronics and Automation,* pp. 1–20. New York: Routledge.

Schmitz, Hubert, and Jose Cassiolato, ed. 1992b. *Hi-Tech for Industrial Development.* New York: Routledge.

Schott, Jeffrey J. 1991. "The Global Trade Negotiations: What Can Be Achieved?" Policy Analyses in International Economics 29. International Institute of Economics, Washington, DC.

Schreiber, William. 1988. *The Open-Architecture Receiver.* Cambridge: MIT Media Lab.

Schreiber, William. 1990. "The Economics and Politics of High-Definition Television." unpublished manuscript. Massachusetts Institute of Technology, March.

Schrieber, William. 1993. "Advanced Television in the United States." report to the European Commission. Massachusetts Institute of Technology, July.

Schreiber, William F., and Robert Greenberg. 1991. "High Definition TV: American Competition in the Twenty-First Century." The Federalist Research Institute.

Schumpeter, Joseph A. 1939. *Business Cycles: A Theoretical, Historical and Statistical Analysis of the Capitalist Process.* New York: McGraw Hill.

Schumpeter, Joseph A. 1942. *Capitalism, Socialism, and Democracy.* New York: Harper and Row.

Schwartz, Herman M. 1994. *States Versus Markets: History, Geography, and the Development of the International Political Economy.* New York: St. Martin's.

Schwoch, James. 1990. *The American Radio Industry and Its Latin American Activities, 1900–1939.* Urbana: University of Illinois Press.

Scott Morton, Michael S., ed. 1989. *Management in the 1990s Research Program Final Report.* Cambridge: MIT Press.

Seib, Gerald F., and David Wessel. 1990. "Three Free-Marketeers Shape Bush's Domestic and Economic Policy." *Wall Street Journal.* April 27.

Setzer, Florence, and Jonathan Levy. 1991. "Broadcast Television in a Multichannel Marketplace." June. OPP Working Paper Series no. 26, Federal Communications Commission Office of Plans and Policy.

Shannon, Claude E., and Warren Weaver. 1953. *The Mathematical Theory of Communication.* Urbana: University of Illinois.

Shew, William. 1993. "Information Superhighways: Is There a Role for Government?" *The American Enterprise* (July August):18–24.

Shooshan, Harry M., III, ed. 1984. *Disconnecting Bell: The Impact of the AT&T Divestiture.* New York: Pergamon Press.

Sinclair, Andrew. 1981. *Corsair: The Life of J. Pierpont Morgan.* Boston: Little Brown.

Sinha, Inkhil. 1993. "Technological Unemployment in the Information Age." International Communication Association Annual Conference. Washington, DC.

Sirbu, Marvin A. 1989. "Telecommunications Standards, Innovation and Industry Structure." February. Carnegie Mellon University.

Sirbu, Marvin A. 1992. "The Struggle for Control Within the Telecommunications Networks," in Sapolsky, Harvey, W. Russell Neuman, Eli Noam, and Rhonda Crane, ed., *The Telecommunications Revolution: Past, Present and Future,* pp. 140–148. New York: Routledge.

Sirbu, Marvin. 1993. "Wireline Networks." National Research Council.

Smith, Adam. 1993 [1776]. *An Inquiry into the Nature and Causes of the Wealth of Nations.* New York: Oxford University Press.

Smith, George D. 1985. *The Anatomy of a Business Strategy: Bell, Western Electric and the Origins of the American Telephone Industry.* Baltimore: Johns Hopkins University Press.

Smith, George D., and D. C. Pitt. 1991. "Open Network Architecture: Journey to an Unknown Destination." *Telecommunications Policy* 15(5):379–394.

Society of Motion Picture and Television Engineers. 1992. "Report of the Task Force on Headers/Descriptors." *SMPTE Journal* (June).

Snyder, Samuel S. 1980. "Computer Advances Pioneered by Cryptologic Organizations." *Annals of the History of Computers* 2 (January): 60–70.

Solomon, Richard Jay. 1977. "The Consumer-Communications Reform Act: A Case of Déja Vù." *Telecommunications Policy* 1(4):347–348.

Solomon, Richard Jay. 1978. "What Happened After Bell Spilled the Acid?" *Telecommunications Policy.* 2(2):146–157.

Solomon, Richard Jay. 1987a. "Electronic and Computer-Aided Publishing." Ad-Hoc Expert Meeting on Information Technologies and Emerging Growth Areas, Organization for Economic Cooperation and Development, Paris.

Solomon, Richard Jay. 1987b. *Open Network Architectures and Broadband ISDN: The Joker in the Regulatory Deck*. ICCC-ISDN'87—Evolving to ISDN in North America. Dallas: IEEE.

Solomon, Richard Jay and Loretta Anania. 1988a. "Black Boxes in the Piazza." *Communications Week International* (September 15).

Solomon, Richard Jay. 1988b. "Vanishing Intellectual Boundaries: Virtual Networking and the Loss of Sovereignty and Control." *Annals of American Political Science* 495 (January).

Solomon, Richard Jay. 1989a. "The Demand for High-Resolution Systems." August 1. Testimony before the Committee on Governmental Affairs, United States Senate, Hearings on the Scope of the HDTV Market and Its Implications for Competitiveness.

Solomon, Richard Jay. 1989b. "RCA and the Origins of Broadcasting." DOHRS-TR-7. February. MIT.

Solomon, Richard Jay. 1989c. "The Role of Government in HDTV Production Standards." May 31. Testimony before the Subcommittee on International Scientific Cooperation, Committee on Science, Space and Technology, U.S. House of Representatives, Hearings on the Impact of International Standards on High Definition Television, Program on Communications Policy and The MIT Media Laboratory.

Solomon, Richard Jay. 1989d. "Shifting the Locus of Control: Computers and Communication in the 1990s," in ed., *Paradigms Revisited*, pp. 1–44. Queenstown MD: Aspen Institute.

Solomon, Richard Jay. 1990a. "Broadband Communications as a Development Problem." *STI Review* 7:65–100.

Solomon, Richard Jay. 1990b. "The Economic Dimension of Standards in Information Technology: New Paradigms for Future Standards." DSTI/ICCP/EIIT/90.4. July. Paris: Organization for Economic Cooperation and Development Directorate for Science, Technology and Industry Committee for Information, Computer and Communications Policy.

Solomon, Richard Jay. 1990c. "HDTV: Digital Technology's Moving Target." *Intermedia* 18(2):58–61.

Solomon, Richard Jay. 1991. "Past and Future Perspectives on Communications Infrastructure," in Elton, Martin C. J., ed., *Integrated Broadband Networks: The Public Policy Issues*, Amsterdam: North-Holland.

Solomon, Richard Jay, and Branko Gerovac. 1994. "Protect Revenues, Not Bits." MIT/Harvard Interactive Multimedia Association Workshop on Intellectual Property, June 1993. Cambridge.

Solow, Robert M. 1959. "Investment and Technical Progress," in Arrow, K. J., Karlin, S., and Suppes, P., ed., *Mathematical Methods in the Social Sciences*, Stanford: Stanford University Press.

Staelin, David. 1991. "High Definition Systems: US Infrastructure and Competitiveness." May 14. Subcommittee on Technology and Competitiveness, Committee on Science, Space and Technology, U.S. House of Representatives.

Stallings, Richard. 1987. *Handbook of Computer Communications Standards.* New York: MacMillan.

Starling, Grover. 1983. "Technological Innovation in the Communications Industry: An Analysis of the Government's Role," in Havick, John J., ed., *Communications Policy and the Political Process,* pp. 171–201. Westport, CT: Greenwood Press.

Stehman, J. Warren. 1925. *The Financial History of the American Telephone and Telegraph Company.* Boston: Houghton Mifflin.

Stigler, George J. 1971. "The Theory of Economic Regulation." *Bell Journal of Economics and Management Science* 2:3–21.

Stigler, George J. 1975. *The Citizen and the State: Essays on Regulation.* Chicago: University of Chicago Press.

Stone, Alan. 1991. *Public Service Liberalism: Telecommunications and Transitions in Public Policy.* Princeton: Princeton University Press.

Strassmann, Paul A. 1985. *Information Payoff.* New York: Free Press.

Strassmann, Paul A. 1990. *The Business Value of Computers.* New Canaan, CT: Information Economics Press.

Strassmann, Paul A. 1995. *The Politics of Information Management: Policy Guidelines.* New Canaan, CT: Information Economics Press.

Subcommittee on Telecommunications, and Finance, Committee on Energy and Commerce, United States House of Representatives. 1987. "Hearing on Telecommunications Trade." Subcommittee on Telecommunications and Finance, Committee on Energy and Commerce, U.S. House of Representatives.

Subcommittee on Telecommunications, Consumer Protection and Finance, Committee on Energy and Commerce, U.S. House of Representatives. 1981. *Hearings: Status of Competition and Deregulation in the Telecommunications Industry.* Washington, DC: Government Printing Office.

Sullivan, Thomas E., ed. 1991. *The Political Economy of the Sherman Act: The First One Hundred Years.* New York: Oxford University Press.

Summers, Mark Wahlgren. 1993. *The Era of Good Stealings.* New York: Oxford University Press.

Taylor, George R. 1951. *The Transportation Revolution, 1815–1860.* New York: Holt Rinehart Winston.

Taylor, George R. and Irene Neu. 1956. *The American Railway Network, 1861–1890.* Cambridge: Harvard University Press.

Temin, Peter. 1987. *The Fall of the Bell System: A Study in Prices and Policies.* New York: Cambridge University Press.

Temin, Peter. 1990. "Cross Subsidies in the Telephone Network after Divestiture." *Journal of Regulatory Economics* (2):349–362.

Teske, Paul, and John Gebosky. 1991. "Local Telecommunications Competitors: Strategy and Policy." *Telecommunications Policy* 15(5):429–436.

Thompson, R. 1947. *Wiring A Continent: The History of the Telegraph Industry in the United States, 1832–1866*. Princeton: Princeton University Press.

Thor, G.M. 1990. "Getting the Most from Productivity Statistics." *National Productivity Review* 9(4):457–466.

Thorelli, Hans Birger. 1954. *The Federal Antitrust Policy: Origination of an American Tradition*. London: Allen and Unwin.

Thurow, Lester C. 1981. "Solving the Productivity Problem," in Thurow, Lester C., Packer, A., and Samuels, H.J., ed., *Strengthening the Economy: Studies in Productivity*, Washingon, DC: Center for Democratic Policy.

Thurow, Lester C. 1985. *The Zero-Sum Solution*. New York: Simon and Schuster.

Thurow, Lester C. 1992. "Is Telecommunications Really Revolutionary?" in Sapolsky, Harvey, W. Russell Neuman, Eli Noam, and Rhonda Crane, ed., *The Telecommunications Revolution: Past, Present and Future*, pp. 1–5. New York: Routledge.

Titch, Steven, and Charles F. Mason. 1992. "Digital Cellular: What Now?" *Telephony* (February 10):30–36.

Tuchman, Barbara W. 1958. *The Zimmerman Telegram*. New York: Viking.

Tunstall, Jeremy. 1986. *Communications Deregulation: The Unleashing of America's Communications Industry*. New York: Basil Blackwell.

Tyler, Michael, ed. 1988. *Economic Factors and Market Penetration*. Washington, DC: Federal Communications Commission Advisory Committee on Advanced Television Services, Planning Subcommittee, Working Party 5.

Tyson, Laura D'Andrea. 1992. *Who's Bashing Whom? Trade Conflict in High-Technology Industries*. Washington, DC: Institute for International Economics.

Ungerer, Herbert. 1990. "Telecommunications Europe." Commission of the European Communities.

United States Department of Commerce. 1990a. *Emerging Technologies: A Survey of Technical and Ecomonic Opportunities*. Washington, DC.

United States Department of Commerce. 1990b. *U.S. Telecommunications in a Global Economy: Competitiveness at a Crossroads*. Washington, DC.

United States Senate. 1990. *Telecommunications Equipment Research and Manufacturing Competition Act: Hearings before the Subcommittee on Communications of the Committee on Commerce, Science and Transportation*. Washington, DC: Government Printing Office.

van Cuilenburg, Jan, and Paul Slaa. 1994. "Competition in the Local Loop: Towards an Anticyclical Competition Policy in Telecommunications." *Telecommunications Policy* 18(1):51–65.

Veblen, Thorstein. 1978. *The Theory of Business Enterprise*. New Brunswick, NJ: Transaction.

Vickers, John, and George Yarrow. 1989. *Privatization: An Economic Analysis.* Cambridge: MIT Press.

Vogel, D. 1978. "Why Businessmen Distrust Their State: The Political Consciousness of American Corporate Executives." *British Journal of Political Science* 8:45–77.

Waldrop, Frank C. 1938. *Television: A Struggle for Power.* New York: Morrow.

Walker. 1939. "Investigation of the Telephone Industry in the United States." House Document No. 340, 76th Congress, 1st Session. Report to Congress, Federal Communications Commission.

Walton, Gary M., and Hugh Rockoff. 1994. *History of the American Economy.* Fort Worth, TX: Dryden.

Ward, James A. 1986. *Railroads and the Character of America, 1820–1887.* Knoxville: University of Tennessee Press.

Weaver, Warren. 1949. "The Mathematics of Communication." *Scientific American.* 181:11–15.

Webre, Philip. 1988. "Federal Financial Support for High Technology Industries." Congressional Budget Office.

Weinberg, Steve. 1983. "The Politics of Rewriting the Federal Communications Act," in Havick, John J., ed., *Communications Policy and the Political Process,* pp. 71–87. Westport, CT: Greenwood Press.

Weinhaus, Carol, Sandra Makeeff, and Peter Copeland. 1994. "Redefining Universal Service: The Cost of Mandating the Deployment of New Technology in Rural Areas." Telecommunications Industries Analysis Project.

Weizenbaum, Joseph. 1976. *Computer Power and Human Reason.* San Francisco: W. H. Freeman.

Weller, Timothy N., and Seema R. Hingorani. 1994. "Information Superhighway: Putting the Pieces Together." Donaldson, Lufkin, and Jenrette.

Wenders, John T. 1987. *The Economics of Telecommunications.* Cambridge: Ballinger.

West, Nigel. 1986. *GCHQ: The Secret Wireless War, 1900–86.* London: Weidenfeld and Nicolson.

Westmeyer, Russell E. 1952. *Economics of Transportation.* Englewood Cliffs, NJ: Prentice-Hall.

Wigand, Rolf T. 1988. "Integrated Services Digital Networks: Concepts Policies and Emerging Issues." *Journal of Communication* 38(1):29–49.

Wildavsky, Aaron. 1986. "Industrial Policies in American Political Cultures," in Barfield, Claude E., and William A. Schambra, ed., *The Politics of Industrial Policy,* pp. 15–32. Washington, DC: American Enterprise Institute.

Williams, Frederick. 1991. *The New Telecommunications: Infrastructure for the Information Age.* New York: Free Press.

Williams, Frederick, and Susan Hadden. 1993. "On the Prospects for Redefining Universal Service: From Connectivity to Content," in Ruben, Brent, and Jorge Schement, ed., *Between Communication and Information*, pp. 401–419. New Brunswick, NJ: Transaction Books.

Williamson, Oliver E. 1985. *The Economic Institutions of Capitalism: Firms, Markets, Relational Contracting*. New York: Free Press.

Wilson, James Q., ed. 1980. *The Politics of Regulation*. New York: Basic Books.

Winkler, John K. 1930. *Morgan the Magnificent*. Babson Park: Spear and Staff.

Witte, Eberhard. 1988. *Restructuring of the Telecommunications System: Report of the Government Commission for Telecommunications*. Heidelberg: R. v. Decker's Verlag, G. Schenck.

Wolf, Charles, Jr. 1988. *Markets or Governments: Choosing Between Imperfect Alternatives*. Cambridge: MIT Press.

Womack, James P., Daniel T. Jones, and Daniel Roos. 1990. *The Machine that Changed the World*. New York: Macmillan.

Wriston, Walter B. 1992. *The Twilight of Sovereignty*. New York: Scribner's.

Yoder, Stephen Kreider, and G. Pascal Zachary. 1993. "Digital Media Business Takes Form as a Ballet of Complex Alliances." *Wall Street Journal*. June 14.

Zachary, G. Pascal. 1991. "Apple Wants to Grow Far Beyond Its Core." *Wall Street Journal*. September 27.

Zorpette, G. 1989. "Keeping the Phone Lines Open." *IEEE Spectrum* (June): 32–27.

Zuboff, Shoshana. 1988. *In the Age of the Smart Machine*. New York. Basic Books.

Zysman, John. 1983. *Governments, Markets, and Growth: Financial Systems and the Politics of Industrial Change*. Ithaca: Cornell University Press.

Index

Above 890 decision, 175, 194
Abrams v. the United States, 203
ACATS, 215–216, 220
ACTS, 15–16, 149–150
Advanced Communications Technologies and Services (ACTS), 15–16, 149–150
Advanced Intelligent Network (AIN) proposals, 105
Advanced Research Projects Agency, 75, 107, 188–189, 217–219
Advanced Technology Program (ATP), 172, 217–218
Advanced television (ATV) material, 215
Advanced Television Research Program, xii
Advanced Television Systems Committee (ATSC), 211–212, 219
Advisory Committee on Advanced Television Services (ACATS), 215–216, 220
AEA, 218
AFJ, 167
AIN proposals, 105
Air defense system, 70–71, 103, 176
Airline reservations systems, 75
All-Channel Receiver Act of 1958, 208–209
Allen, Robert, xv
AM radio transmission, 205, 236
American Electronics Association (AEA), 218

American Newspaper Publishers Association, 12, 190–191
American Telephone and Telegraph. *See* AT&T
American Telephone Company, 52
Amplitude modulation (AM) radio transmission, 205, 236
Analog networks, 67–70
Antitrust legislation
  AT&T and, 5, 161, 163, 174, 261
  computers and, 228
  future of, 174–175
  IBM and, 261
  judiciary's role in, 170–171
  railroad and, 5, 53
  regulatory policy and, 240
  transportation networks and, 5
  Western Union and, 91–92, 186–187
Armstrong, Edwin, 195, 205
ARPAnet
  beginning of, 75
  EUREKA program and, 148
  high-definition television and, 217–219
  Internet and, 188–189
  packet networks and, 75, 104, 188
  research and, 188–189, 222
  standards and, 76, 107
Articles of Confederation, 87
Artificial intelligence, 228, 232
ASCII, 234
Association of Maximum Service Telecasters, 214

Asymmetry
  of established player and newcomer, 245
  of regulator and industry, 245–246
  of vendor and customer, 245
Atomic Energy Commission, 70, 103, 226
ATP, 172, 217–218
ATSC, 211–212, 219
AT&T
  Above 890 decision and, 175
  antitrust legislation and, 5, 161, 163, 174, 261
  Bell Laboratories and, 70, 156, 231
  cellular licenses and, 236
  complaints against, 1910–1913, 161–167
  computers and, 184–185, 229–231
  Consent Decree and, 71, 184
  cream-skimming accusations and, 178–179, 183
  development of, 156–161
  divestiture of, 174, 191–193, 204, 235
  enhanced services of, 189–191
  foreign attachments and, 176–178
  Hi-Lo tariff and, 179
  IBM and, 229
  indictment of, 169
  Kingsbury Commitment and, 55, 165, 173
  lobbying by, 182
  Long Lines and, 55, 156, 158, 191
  National Information Infrastructure and, 199
  origins of, 52, 54–55
  Pulse Code Modulation and, 68
  radio and, 196–199, 201–202
  regulatory policy and
    "harm to networks" and, 176–178
  MCI and, 178–180
  rates, 11
  rights-of-way and, 175
  satellite communications and, 174
  telegraph and, 52
  telephone patents of, 52–55, 159

  Telpak tariffs and, 175
  Western Electric and, 158, 177
  Western Union and, 157–159, 164, 250

Babbage, Charles, 101
Baudot, Emile, 74–75, 99
Baudot Teletype, 74–75, 99, 223
Bell, Alexander Graham, 52
Bell Laboratories, 70, 156, 231
Bell Operating Companies (BOCs), 12, 19, 191–193, 235–236
Bell System Crossbar units, 105
Bell Telephone model, 100–101
Bicycles, 92
Bill of Rights, 202
Bingaman, Anne, 173, 260
Blue network, 198–199
BOCs, 12, 19, 191–193, 235–236
Born to Lead (Nye), 16
Brandeis (Justice), 203, 206
Broadcasting. *See also* Radio; Television
  amplitude modulation transmission, 205, 236
  frequency modulation transmission, 205, 236
  Kingsbury Commitment and, 201
  National Information Infrastructure and, 209
  private, 27
  regulatory policy, 9–11
  Russian, 194
  wave, 205
Broadcasting wars, 194–223
  Advanced Research Projects Agency and, 217–219
  Advisory Committee on Advanced Television Services and, 215–216
  All-Channel Receiver Act of 1958 and, 208–209
  alternative systems proposals and, 216
  background, 194–195
  cable television and, 209–211
  color television and, 205–208

Committee for Open High Resolution Systems and, 219–223
Communications Act of 1934 and, 200–206
Congress and, 216–217
Dubrovnik meeting and, 213–214
Federal Communications Commission and, 214–215
high-definition television and, 211–223
Open Communications Infrastructure and, 205
RCA and, 195–200
Brown, George, 217
Brown network, 198
Brownell, Herbert, 174
Brunel, Isambard Kingdom, 97
Brynjolfsson, Erik, 119–121, 134
Bullard (Admiral), 195
Bureau of Communication and Information Policy, 212, 214
Burleson (Postmaster), 165
Bush administration, 18, 21, 138, 221
Bush, George, 217–218

Cable Act of 1992, 11, 254–255
Cable television, 10–11, 209–211
Cable TV Act of 1984, 11, 240
Calculator, pocket, 104
Canal building, 89–90
Capitalism, 5, 115
CAPTAIN videotext system, 143
Carter administration, 168, 172, 192
Carterphone case, 177–178, 188, 192
Case studies. *See also specific companies*
political economics, 23–25
purpose of reviewing, xviii
Cathode ray tube (CRT), 103, 232
CCIR, 196, 207, 211, 213, 218–219
CDC, 229
CEI, 110
Cellular licenses, 236–237
Cellular technology, 29, 237–239, 242
Civil Aeronautics Act of 1938, 249
Civil Aeronautics Board, 260

Clayton Act of 1914, 5, 164, 204
Clinton administration, 22–23, 39, 138, 259–260
Clinton, Bill, xv, 18, 173
Closed models, 95–98
Code-division multiplexing, 241–242
COHRS, 139–140, 218–223
Coll, Steven, 174
Color television, 205–208
Colossi, 229
Comité Consultatif International des Radio Communications (CCIR), 196, 207, 211, 213, 218–219
Committee for Open High Resolution Systems (COHRS), 139–140, 218–223
Common carriage model, 27, 61, 64, 88–90
Common Carrier Bureau, 188
Common-carrier law, 64
Communications Act of 1934
amendments to, 171
broadcasting wars and, 200–206
communications technology and, 240, 248
computers and, 104
foreign ownership and, 258
Interstate Commerce Act of 1887 and, 63, 94
political gridlock in communications technology and, 55
public-trustee model and, 200–206
Communications policy
evolution of, 18–20
paralysis in, 20–23
Communications Satellite Corporation (COMSAT), 174
Communications technology. *See also* Political gridlock in communications technology; *specific types*
adaptation of, into human daily ritual, 1–2
changes resulting from, xi–xii
Communications Act of 1934 and, 240, 248
competition created by, 2–3, 12–14

Communications technology (cont.)
  computers and, 105–108
  digital revolution in, xi, xiii, 2, 14,
    58, 64–66
  economy and, U.S., 3, 58, 116
  in Europe, 14–16
  in Japan, 14–16
  models for regulation in, 26–27
  policy strategy for managing, xiii
  politics and, 3
  predictions about, 43–44
  schools of thought in regulating, xiii–
    xiv
  technical implosion caused by, 2
  time line, 81–84
  transportation networks and, xv, 59–
    63
Communications theory, 66
Compact disc, 2
Comparatively Efficient Interconnec-
    tion (CEI), 110
Competition, global, xvii, 14–18,
    126–127, 134–135
Computer Inquiry I proceedings, 71,
    187–189, 194
Computer Inquiry II proceedings, 71,
    190–191, 194
Computer Inquiry III proceedings, 71,
    194
Computers
  antitrust legislation and, 228
  artificial intelligence and, 228, 232
  AT&T and, 184–185, 229–231
  Bell Crossbar units, 105
  bits and, 73–74
  characteristics of, 65–70
  as communicating devices, 80
  communication and, 184–185
  Communications Act of 1934 and,
    184
  communications technology and,
    105–108
  conversion, EBCDIC to ASCII, 234
  copyright legislation and, 11
  digital, 101–105

  digital revolution and, 69–70
  electronic digital, 229
  Federal Communications Commis-
    sion investigation of, 71
  government investment in, 102–103
  IBM 701, 226
  IBM and, 103, 186, 223–232
  interface standards for, 75
  mainframe, 231–232
  memory chips and, 17
  microelectronics and, 72–73
  networked education systems and,
    131
  packet networks for, 75
  productivity and, 118–119
  regulatory policy, 228–235
  semiconductors and, 17, 141
  stored-program control, 71, 76, 102,
    105, 185, 230
  time-sharing, 185–191, 233
  as utility, 233–234
  Whirlwind, 228, 232
COMSAT, 174
Congress, U.S., 171–172, 208, 216–
    217, 259–260
Connectivity, 55
Consent Decree of 1956, 71, 173–174,
    184–185
Constitution, U.S., 87, 171, 208
Consultative Committee for Interna-
    tional Radio (CCIR), 196, 207,
    211, 213, 218–219
Control Data, 103
Control Data Corporation (CDC), 229
Control paradigms, 47
Copyright legislation, 11
Crane, Rhonda, 206–207
Cream-skimming, 178–179, 183
CRT, 103, 232
Cullom Act of 1887, 5, 63, 92, 94,
    201, 204, 248
Cultural lags, 227–228, 244
Cummings, Homer, 164
Cyberspace, 4, 63
Cyclades network, 75, 107

Data transmission standards, 98–100
Data under voice (DUV), technology, 180
Dataphone digital service, 180
Date certain language, 256
Datran Corporation, 179–180
Davidson, William, 19–20
Davidson-Selwyn debates, 19–20
DDS service, 181
de Gaulle, Charles, 206
DEC, 72, 103, 227
Declinists, 16–18
DeForest's Audion amplifier vacuum tube patents, 60
Denver Railroad, 48–49, 51
Department of Commerce, 172, 217–218
Department of Defense (DoD), 72, 75, 104, 107, 188, 217–219
Department of Energy (DoE), 72
Deregulation, 167–168, 170–171
Digital electronic networks. *See* Networks
Digital Equipment Corporation (DEC), 72, 103, 227
Digital networks, 67–70
Digital revolution
  in communications technology, xi, xiii, 2, 14, 58, 64–66
  computers and, 69–70
  networks and, 64–66
  wireless technology and, 241
Direct Distance Dialing, 104
Directorate General 13, 148, 213
Dissimulation, 245
DoD, 72, 75, 104, 107, 188, 217–219
DoE, 72
Domestic Communications Satellite Facilities decision, 179
Dornbusch, Rudiger, 126–127
Dougan, Diana, 213
DRAM semiconductor manufacturing, 141
Draper Laboratories, 104
Drucker, Peter F., 115

Dual-use technology, 104–105
Dubrovnik meeting, 213–214
DUV technology, 180
Dynamic random access memory (DRAM) semiconductor manufacturing, 141

EBCDIC character set, 234
Economic conflict, historical precedent in, 4–5
Economy
  communications technology and, 3, 58, 116
  political, 23–25
  political gridlock in communications technology and, 115
  United States, xv-xvi, 3, 58, 116, 135–141, 146
Electrical Products Research Incorporated (EPRI), 184
Electronic digital computers, 229
Electronic mail, 63
Electronic marketplace, 114–115
Electronic Research Associates (ERA), 103, 229
ENFIA, 183
Eniac, 229
EPRI, 184
Equal time provision, 203
ERA, 103, 229
Erie Canal, 89
ESPRIT, 15, 149–150
EU, 15, 133–134, 147
EUREKA program, 16, 148–150
Europe
  cellular technology in, 242
  color television in, 206
  communications technology in, 14–16
  high-definition television in, 148–149, 221–222
  industrial policy of, 21
  manufacturing in, 17, 128
  networked education system in, 129
  telecommunications industry in, 146–151

Europe (cont.)
  telecommunications research projects
    in, 140–141
  telematics policy in, 146–150
Europe 1992 program, 146
European Strategic Programme for
  Research in Information Tech-
  nology (ESPRIT), 15, 149–150
European Union (EU), 15, 133–134,
  147
Everett (Governor), 46–47
Exchange network facilities for inter-
  state access (ENFIA), 183
Execunet case, 180–183
Executive branch as player in tele-
  communications wars, 172–175
Exon (Senator), 203

Faxes, 63, 69
FCC. *See* Federal Communications
  Commission
Federal Communications Act of 1934.
  See Communications Act of 1934
Federal Communications Commission
  (FCC)
  Advanced Television standards-
    setting process and, 222
  broadcasting wars and, 214–215
  cable television and, 10, 209–210
  computer investigation by, 71
  deliberations of, 172
  establishment of, 166, 200
  high-definition television and, xiv,
    214–215
  industrial policy and, 20
  MCI and, 179
  personal communications systems
    and, 239
  as player in telecommunications wars,
    168–170
  Report and Order on Open Network
    Architecture and, 110
  specialized common carrier and, 180
  standards and, 139
  Telecommunications Act of 1996
    and, 31–32
  Telpak tariffs and, 175

Federal policy
  government's role in, 85–88
  public interest and, 88–89
  themes in, historical, 85–86
Federal Radio Act of 1912, 248
Federal Radio Act of 1927, 198, 200,
  203–204, 248
Federal Radio Commission, 200
Federal Trade Commission, 164
Fiber-to-the-home (FTTH) project,
  144
Fields, Craig, 222
Final Judgment of 1956, 55, 173, 185
First Amendment, 202–203, 208–209
Flexible access, 30, 255
Flowers, Thomas, 71, 102
FM radio transmission, 205, 236
Forbes family, 53–55
Ford, Henry, 158
Foreign investors in United States,
  136–137
Forrester, Jay, 104, 228
Fourth National Radio Conference,
  202
France, 98, 206–207
Frequency modulation (FM) radio
  transmission, 205, 236
"Friendly family" of standards, 216
FTTH project, 144
Fulton, Robert, 87

Gallatin, Albert, 89
GATT Uruguay Round, 126, 133–134
GE, 195–198
General Agreement on Tariffs and
  Trade (GATT) Uruguay Round,
  126, 133–134
General Electric (GE) Company, 195–
  198
General Instruments (GI), xiv–xv, 221
General Telephone and Electric (GTE),
  189, 261
Germany, 128–129
GI, xiv–xv, 221
Gibson v. Ogden (1924), 87
Global competition, xvii, 14–18, 126–
  127, 134–135

Global Information Infrastructure, xv, 39–40, 42, 114, 140, 151
Good Roads Movement, 92–93, 158
Gordian Knot
  argument, 260
  cutting, xviii, 167, 247–250
  legend of, xi
  Open Communications Infrastructure and, 167
  Report and Order on Open Network Architecture and, 110
  stages of telecommunications development and, 156
  usage today, xi
Gore, Al, xv, 18, 22, 173
Gould, Jay, 52–54
Government Services Administration, 186
Great Debate, xii–xiii, xv
Great Western Railway, 97
Groupe Bull, 150
GTE, 189, 261

HD-MAC system, 145, 148
HDTV. *See* High-definition television
Header group, 220
Heliocentric model of planetary motion, 7
Hendrick, Burton, 87–88
Hierarchical model, 73–74, 78–80
High Performance Computing and Communication (HPCC) program, 22
High-definition television (HDTV)
  ARPAnet and, 217–219
  broadcasting wars and, 211–223
  competition in, xii–xiv
  Department of Commerce and, 217
  in Europe, 148–149, 221–222
  Federal Communications Commission and, xiv, 214–215
  in Japan, xiv, 143, 145, 148, 221–222
  legislation supporting, 217
  paradox of, 211–213

Hi-Lo tariff, 179
Hirschman, Albert, 257
Hitt, Lorrin, 119–121
Hi-Vision, xiv, 143
Hollerith, Herman, 101
Holmes (Justice), 203, 206
Horwitz, Robert Brett, 176
House Telecommunications Subcommittee, 216–217
HPCC program, 22
Hush-A-Phone, 176

IBM
  air defense system and, 70
  antitrust legislation and, 261
  AT&T and, 229
  computers and, 103, 186, 223–232
  layered architecture and, 234
  origins of, 101
  tabulations and, 101
  time-sharing and, 185–186
IBM 701, 226
ICC, xvi, 53–54, 161, 163, 200
IEEE-USA, 219–220
Illinois Central Land Grant of 1850, 91
Income, American, 17, 126
Industrial policy, 20–24
Industrial Revolution, 154
Information highway metaphor, xv, 22
Information network system (INS), 15
Information processing, 122
Information Superhighway metaphor, 22
Information technology. *See* Communications technology
Infrastructure models, 94–101
  Bell model, 100–101
  closed, 95–98
  telegraph and data transmission standards, 98–100
  transport, 94–95
Innovation in Communication and Information Technologies, 148, 213
INS, 15

Institut de Recherche d'Information et d'Automatique (IRIA), 75, 107
Institute of Electrical and Electronic Engineers (IEEE-USA), 219–220
Integrated Services Digital Networks (ISDN), 75, 105, 107, 142–143, 145
Intelligence quotient (IQ) measures, 129
Interconnection, 259
International Business Machines. *See* IBM
International Mobile Machines, 241
International Standards Organization (OSI), 108, 220–221
International Telecommunication Union, 99, 196, 220–221
International Telegraph Union, 99
Internet
  ARPAnet and, 188–189
  censorship and, 43
  commercial enterprises and, 138–140
  development of, 139
  internetworking and, 66
  Japanese, 145
  networks and, 79
  Open Communications Infrastructure and, 80–81
  physical versus logical networks and, 62
  in United States, 145
  unpredictability of, 40–41
Internet Engineering Task Force, 220–221
Internet Protocol (IP), 43, 76
Internetworking, 66, 78
Interstate and Defense Highway Act of 1956, 22, 93
Interstate Commerce Act of 1887, 5, 63, 92, 94, 201, 204, 248
Interstate Commerce Commission (ICC), xvi, 53–54, 161, 163, 200
IP, 43, 76
IQ measures, 129
IRIA, 75, 107
ISDN, 75, 105, 107, 142–143, 145
ITU-R, 196

Jackson, Charles L., 236
Jacquard punched cards, 101
Japan
  communications technology in, 14–16
  computer communications services in, 144–145
  high-definition television in, xiv, 143, 145, 148, 221–222
  industrial policy of, 21
  information society in, 140–141
  Internet in, 145
  manufacturing in, 17, 128
  network industrial organization in, 123
  network policy in, 141–146
  networked education system in, 129
  Open Communications Infrastructure in, 146
  railroad in, 97–98
  regional informatization programs in, 143
  telecommunications industry in, 141–146, 151
  telecommunications research projects in, 140–141
  Teletopia large-scale broadband pilot projects in, 143
  zaibatsu tradition in, 21
Jonscher, Charles, 121–122, 127
Judiciary as player in telecommunications wars, 170–171

Kahn, Alfred, 168
Kavner, Robert, 13
KDKA radio station, 197
Kelly, Tracey E., 236
Kindleberger, Charles, 154
Kingsbury Commitment of 1913
  AT&T and, 55, 165, 173
  broadcasting and, 201
  political gridlock in communications technology and, 55
  regulatory policy and, 63, 160, 171, 248
  telecommunications industry and, 63, 160, 248

telephone and, 55, 165, 173
universal service and, 204
Willis-Graham Act and, 204
Kingsbury, N. C., 160, 164–165
Krugman, Paul, 17, 113, 127

Laissez-faire model, 18, 28, 64
Land-Grant Telegraph Act of 1861, 91–92
LANs, 75, 107
Lippman, Andrew, 216
Local area networks (LANs), 75, 107
Logical networks, 58, 62, 80–84
Long Lines, 55, 156, 158, 191
Lovelace, Ada, 101

McCarthy, John, 233
McGowan, William, 178
Machlup, Fritz, 122
MCI, 178–182, 191
McReynolds, James C., 160, 164
Mail, 63, 69
Mainframe computers, 231–232
Malone, Thomas W., 124–125, 134
Mann-Elkins Act of 1910, 3, 161, 179, 200, 204, 248
Manufacturing
  dynamic random access memory
  in Europe, 17, 128
  in Japan, 17
  Open Communications Infrastructure and, 128–129
  productivity and networks and, 125–129
  semiconductor, 141
  in United States, 16–17, 128
Marconi Wireless Telegraph Company, 195–196
Marketplace, 58–59, 114–115
Markey, Edward, 216–217
Massachusetts Institute of Technology (MIT), xii
Media convergence, 70–72, 76
Memory chips, 17
Merrifield, Bruce, 218

MFJ of 1982, 55, 166–167, 174, 191–193
Microcell technology, 258
Microelectronics, 72–73
Microwave Communications Incorporated (MCI), 178–182, 191
Microwaves, 174–175
Minc, Alain, 146–147
Ministry of International Trade and Industry, Japan (MITI), 15, 134, 142–144, 151, 251
Ministry of Posts and Telecommunications, Japan (MPT), 142–144
MIT, xii
MIT Commission on Industrial Productivity, 17
MITI, 15, 134, 142–144, 151, 251
Mitterand, François, 148
Modified Final Judgment (MFJ) of 1982, 55, 166–167, 174, 191–193
MOF, 142, 144
Monopoly
  infrastructure, 87–88
  natural, 11, 84, 157, 201
  in New Deal era, 87–88
  railroad, 88, 90
  telegraph, 88, 90
Morgan, J. P., 54, 157, 250
Morita, Akio, 125–126
Morse Code, 74, 98
Morse-Vail telegraph system, 98
Mosbacher, Robert, 217–218
MPT, 142–144
MTS, 179
Multiplexed channels, 105–106, 241–242
Munn v. Illinois (1877), 92, 171, 204
Munro, Richard, 14
MUSE HDTV system, 145, 148
Must carry regulations, 210
Muzak, 197

NAFTA, 126
NASA, 72, 104

National Aeronautics and Space Administration (NASA), 72, 104
National Broadcasting Company (NBC), 197, 199
National Bureau of Standards, 172–173, 217
National Information Infrastructure (NII)
AT&T and, 199
broadcasting and, 209
evolution of communications policy and, 18–19
National Telecommunications and Information Administration and, 172–173
network protocols and, 74
Open Communications Infrastructure and, xv, 140
Telecommunications Act of 1996 and, 40, 42, 262
National Institute of Standards and Technology (NIST), 172–173, 217
National Labor Relations Act of 1935, 249
National Performance Review, 23
National Research and Education Network (NREN) initiative, 22
National Science Foundation, 70
National Security Agency (NSA), 70, 103, 226, 229–230
National Telecommunications and Information Administration (NTIA), 18–19, 172–173
National Television Standards Committee (NTSC), 215
Natural monopoly, 11, 64, 157, 201
NBC, 197, 199
Network industrial organization, 123
Network policy, Japanese, 141–146
Network theory, 78–80
Network wars. *See also* Broadcasting wars; Telecommunications wars
computer, 223–227
pattern in, 155, 243–246
wireless, 235–243

Networked education systems, 129–132
Networked-information technology, 115–125
Networks, 45–84. *See also* Network wars; Productivity and networks; State and networks
analog, 67–70
from convergence to divergence, 80
Cyclades, 75, 107
decentralization, 79
digital, 67–70
digital revolution and, 64–66
Internet and, 79
layers, 76–78
local area networks, 75, 107
logical, 58, 62, 80–84
media convergence and, 70–72, 76
microelectronics, 72–73
nature of, 55–58
packet, 75–76, 104, 188
physical, 58, 62, 80–84, 86
properties of, xvii
protocols for, 74–75, 77–78
railroad, 45–49, 51–54
SAGE, 103–104, 226–228, 230–232
standards, 66–67, 74–75
technology and, 4–5
technology models of, 63–64
telecommunications, 59–63, 65–70
telephone, 52–55
theory, 78–80
time-sharing, 185–191
transportation, 58–63
wideband continental microwave, 227
New Deal initiatives, 93, 168, 224, 249
New Telecommunications Law (1985), 15
New York and Lake Erie Railroad, 97
New York Central and Hudson River Rail Road, 91
Newspaper Association of America, 12, 190–191
Newspapers, 1–2, 8–9, 12

NHK, 143, 145, 211–212, 218, 221
NIFTY Serve, 144
NII. *See* National Information Infrastructure
Nippon Telegraph and Telephone (NTT), 15, 142, 144
NIST, 172–173, 217
Nixon, Richard, 172
NOI, 215
Nora, Simon, 146–147
Nora-Minc Report (1980), 15
North American Free Trade Agreement (NAFTA), 126
Notice of inquiry (NOI), 215
NREN initiative, 22
NSA, 70, 103, 226, 229–230
NTIA, 18–19, 172–173
NTSC-Color system, 206–207, 215, 221
NTT, 15, 142, 144

OCI. *See* Open Communications Infrastructure
OFTEL, 15
Olivetti, 150
Olsen, Ken, 227
Omnibus Trade Act of 1988, 217
ONA, 109–110
Open access, 29, 253–254
Open architecture, 28–29, 252
Open Communications Infrastructure (OCI), 250–264
  broadcasting wars and, 205
  Clinton administration and, 259–260
  Congress and, 259–260
  effect of, 250–251
  elements of, central, 28–30
  future and, 262–264
  general-purpose networks built with, 79–80
  global competition and, 134–135
  Gordian Knot and, cutting of, xviii, 167
  implementation, problems of, 255–259
  Internet and, 80–81

in Japan, 146
manufacturing and, 128–129
National Information Infrastructure and, xv, 140
as natural goal, 251–255
networked education systems and, 130, 132
networked-information technology in, 125
other regulation models and, 26
political gridlock in communications technology and, 3, 25–30, 138
productivity and networks and, 113–114, 150–154
proposition of, xiii
regulatory agencies and, 260–262
state and networks and, 108–111
Telecommunications Act of 1996 and, 262
telecommunications wars and, 193–194, 205
United States economy, future, 135
wireless technology and, 29
Open interconnect, 165
OSI, 108, 220–221

Pacific Telesis (PacTel), 13
Packet Communications Incorporated, 188–189
Packet networks, 75–76, 104, 188
Packet switching, 107
PacTel, 13
PAL color system, 206–207
Palmer, William J., 48–49
Paradigm shift, xiii, 58, 86
Park, Yung Chul, 127
PCM, 68–69
PCN, 29
PCS, 239–243
PC-VAN, 144
Pennsylvania Railroad, 54
Personal communications network (PCN), 29
Personal communications systems (PCS), 239–243
"Petition for Notice of Inquiry," 214

Philips (electronics firm), 148–150, 213
Physical networks, 58, 62, 80–84, 86
Policy. *See also* Communications policy; Federal policy; Regulatory policy
   industrial, 20–24
   network, Japanese, 141–146
Political gridlock in communications technology, 1–44
   Communications Act of 1934 and, 55
   communications policy and evolution of, 18–20
   paralysis in, 20–23
   economic conflict and, historical precedent in, 4–5
   economics and, 115
   global competition and, 14–18
   Kingsbury Commitment and, 55
   Open Communications Infrastructure and, 3, 25–30, 138
   pattern in network wars and, 244
   players in, 5–6
   playing field of, 6–12
   political economics case study and, 23–25
   politics and, 115
   rhetoric of, 12–14
   Telecommunications Act of 1996 and, 30–41, 250
   unintentional consequences and, 41–43
Politics, 3, 115, 203
Pool, Ithiel de Sola, 42, 202–203, 240
Porat, Marc, 122
Postal delivery, 63, 69
Prestowitz, Clyde V., Jr., 17
Price-cap regulation, 170
Private broadcasting systems, 27
Private Sector Vitality Act of 1987, 15
Privatization, 57
Productivity and networks, 113–154
   competitive advantage and, xvii
   in electronic market, 114–115
   European telematics policy and, 146–150

importance of productivity and, 113
   Japanese network policy and, 141–146
   manufacturing and, 125–129
   networked education and productivity and, 129–132
   networked-information technology and firm productivity and, 115–125
   Open Communications Infrastructure and, 113–114, 150–154
   trade and telecommunications and, 132–135
   United States technology policy and telecommunications trade and, 135–141
Productivity growth, 115
Productivity paradox, 119–120
Project MAC, 233
Protocols, network, 74–75, 77–78
Ptolemaic model of heavenly motion, 7–8
Public interest, 88–89, 181, 200–201, 203–204
Public Law 104–104. *See* Telecommunications Act of 1996
Public ownership model, 26–27
Public-trustee model, 27–28, 200–206
Pulse Code Modulation (PCM), 68–69

Quiett, Glenn Chesney, 49, 51–52

RACE program, 16, 149–150
Radio
   adaptation of, into human daily ritual, 1–2
   amplified modulation transmission, 205, 236
   AT&T and, 196–199, 201–202
   Commerce Act of 1887 and, 63
   frequency, modulation transmission, 205, 236
   police, 236
   RCA and, 195–200
   regulatory policy, 9
   telegraphy, 196

telephony, 60
Western Electric and, 198
wireless technology, 60–61
Radio Act of 1912, 248
Radio Act of 1927, 198, 200, 203–204, 248
Radio Corporation of America (RCA), 55, 195–200
Railroad
antitrust legislation and, 5, 53
common-carrier law and, 64
competition in, 158–159
economic power of, 5
economy and, U.S., xvi
in France, 98
government aid to, 90–94
greed and, 47–52
in Japan, 97–98
monopoly, 88, 90
networks, 45–49, 51–54
opposition of, to road building, 93
regulatory policy, 89
standards, 95–98
steam technology and, 88, 95
telecommunications architecture and, 59
telegraph and, 94, 99–100
Ramo, Bunker, 186
RAND Corporation, 107
RBOCs, 12, 19, 191–193, 235–236
RCA, 55, 195–200
Reagan administration, 18, 21, 138, 166
Reagan, Ronald, 192
Red networks, 198–199
Regional Bell Operating Companies (RBOCs), 12, 19, 191–193, 235–236
Regional informatization programs, 143
Regulatory agencies, role of existing, 260–262
Regulatory policy. *See also specific legislation*
antitrust legislation and, 240

AT&T and
"harm to networks" and, 176–178
MCI and, 178–180
rates, 11
background of telecommunications wars and, 156–161
broadcasting, 9–11
computer, 228–235
content, 258–259
deregulation and, 167–168
industrial policy and, 24
of information highway, xv–xvii
Kingsbury Commitment and, 63, 160, 171, 248
must carry, 210
pattern in, 155
players and stakes in, 168–175
Congress, 171–172
executive branch, 172–175
Federal Communications Commission and, 168–170
judiciary branch, 170–171
price-cap, 170
radio, 9
railroad, 89
schools of thought in, xiii, xiv
stages in, 156, 160, 166–167
technology and, 228–235
Telecommunications Act of 1996 and, 22, 38–41
television, 9–11
Reich, Robert, 137, 258
Reno, Janet, 260
Report and Order on Open Network Architecture (ONA), 109–110
Research on Advanced Communication Technologies in Europe (RACE) program, 16, 149–150
Restructuring of the Telecommunications System (1987 report), 15
Rights of way, 175
Rio Grande Railroad, 48–49, 51
Road building, 89–93, 97, 158
Robber barons, 5–6, 47
Rohlfs, Jeffrey H., 236

Roosevelt, Franklin Delano, 93, 164, 168, 195, 197–199, 248
Roosevelt, Theodore, 157, 161

Safeguards, xvii
SAGE network, 103–104, 226–228, 230–232
Sante Fe Railroad, 48
Sarnoff, David, 196
Satellite communications, 174
SCC, 179–180, 188–189
Schivelbusch, Wolfgang, 45–46
Schreiber, William, xiv, 211–212, 216
Schumpeter, Joseph, 115
SECAM, 206–209
Securities Act of 1933, 249
Securities Exchange Act of 1934, 249
Selwyn, Lee, 19–20
Semi-Automatic Ground Environment (SAGE) network, 103–104, 226–228, 230–232
Semiconductors, 17, 141
Shannon, Claude, 70
Sherman Anti-Trust Act of 1890, 5, 91, 161, 204
Sherman, John, 91
Siemens, 150, 213
Signal-to-noise ratio (SNR), 67
Sikes, Alfred, 222
Smith, Adam, 46
SMPTE, 220
Smyth v. Ames (1898), 169
SNA, 234
SNR, 67
Social Security system, 224
Society of Motion Picture and Television Engineers (SMPTE), 220
Solow, Robert, 115
Southwestern Bell, 169
Southwestern Bell Supreme Court decision (1923), 169
SPC, 71, 105, 185, 230
Special Telecommunications Action for Regional Development (STAR), 16
Specialized common carrier (SCC), 179–180, 188–189

Sperry Rand, 229
Staggers Act of 1980, 168
Standard Oil, 161
Standards
advanced television material, 215
ARPAnet and, 76, 107
data transmission, 98–100
Federal Communications Commission and, 139
"friendly family" of, 216
incentives, 246
interface, 75
National Bureau of Standards and, 172–173, 217
network, 66–67, 74–75
open system, 108–109
railroad, 95–98
technical, 259
telegraph, 74–75, 98–100
STAR, 16
Star Wars, 148
State and networks, 85–111
aid to transport and, 90–94
common carriage model and, 88–90
control and, 88–90
convergence of computation and communications and, 105–108
digital computers and, 101–105
infrastructure models and, 94–101
Bell model, 100–101
closed, 95–98
telegraph and data transmission standards, 98–100
transport, 94–95
legislation, 86–88
new paradigm and, 108–110
Open Communications Infrastructure and, 108–111
themes in, historical, 85–86
State Department, 214
Steam technology, 87–88, 95
Stibitz, George, 230–231
Stored-program control (SPC), 71, 76, 102, 105, 185, 230
Strassman, Paul, 118
Strategic Air Defense Command, 231

Strategic Defense Initiative (Star Wars), 148
Supreme Court, U.S., 87, 169. *See also specific cases*
System Network Architecture (SNA), 234

Tabulating Company, 223
Taft, William Howard, 161, 164
TCP, 76
TDM circuits, 178
Technologies of Freedom (Pool), 42
Technology. *See also specific types*
cellular, 29, 237–239, 242
cultural lags and, 227–228, 244
data under voice, 180
debate, xii–xiii, xv
dual-use, 104–105
labor-saving, 122
microcell, 258
models, 63–64
networked-information, 115–125
networks and, 4–5
paradigm shift in, xiii, 58, 86
policy, telecommunications industry and, 135–141
push, 243–244
ramifications of, xi–xii
regulatory policy and, 228–235
steam, 87–88, 95
Telecommunications Act of 1996, 30–41
battle of, 30–31
characterization of, xvi–xvii
content of, xiii
effect of, 31, 41–43
electronic publishing by Bell Operating Companies and, 35–38
Federal Communications Commission and, 31–32
Global Information Infrastructure and, 42
industrial policy and, 22
limitations of, 172
National Information Infrastructure and, 40, 42, 262

Open Communications Infrastructure and, 262
political gridlock in communications technology and, 30–41, 250
regulatory policy and, 22, 38–41
rhetoric of, 31–32
stalemate and, new level of, 20
Title II—Broadcast Services, 33
Title III—Cable Services, 33–34
Title I—Telecommunications Services, 32–33
Title IV—Regulatory Reform, 34
Title VI—Effects on Other Services, 34
Title VII–Other, 34–38
Title V—Obscenity and Violence, 34
Telecommunications industry
competition in, 2 3
connectivity and, 55
in Europe, 146–151
future, 262–264
in Japan, 141–146, 151
Kingsbury Commitment and, 63, 160, 248
research projects, 140–141
stages of development in, 156, 160, 166–167
trade and, 132–135
United States technology policy and, 135–141
Telecommunications network
characteristics of, 65–70
shifts in, 106–107
transportation networks and, 59–63
Telecommunications wars, 156–194
AT&T and
complaints against, 1910–1913, 161–167
development of, 156–161
"harm to networks" and, 176–178
MCI and, 178–180
background of, 156–161
competition and, opening of floodgates, 182–184
computers and communication and, 184–185

Telecommunications wars (cont.)
  deregulation and, 167–168
  Execunet case and, 180–182
  microwaves and, 175
  Modified Final Judgment and, 191–193
  Open Communications Infrastructure and, 193–194, 205
  players and stakes in, 168–175
    Congress, 171–172
    executive branch, 172–175
    Federal Communications Commission, 168–170
    judiciary branch, 170–171
  time-sharing and, computer, 185–191
  wireless, 235–243
Telegraph
  AT&T and, 52
  Baudot Teletype and, 74–75
  control of, 88–89, 91
  introduction of, 4
  Land-Grant Telegraph Act and, 91–92
  monopoly, 88, 90
  Morse Code and, 74
  Morse-Vail system, 98
  railroad and, 94, 99–100
  standards, 74–75, 98–100
  telephone and, 4–5
  Wheatstone-Cooke system, 98
Telematics policy, European, 146–150
Telephone
  adaptation of, into human daily ritual, 1–2
  analog versus digital, 67–70
  AT&T patents for, 52–55, 159
  Bell model, 100–101
  cellular, 29, 237–239
  common carriage model and, 64
  hierarchical model of, 73–74, 78–80
  Kingsbury Commitment and, 55, 165, 173
  media convergence and, 69
  message telephone service, 179
  as natural monopoly, 11

Pulse Code Modulation and, 68–69
  standards, 52–55
  switches, 107–108
  telegraph and, 4–5
  wireless technology and, 60–61
Teletopia large-scale broadband pilot projects, 143
Television. *See also* High-definition television
  adaptation of, into human daily ritual, 1
  cable, 10–11, 209–211
  color, 205–208
  "free" terrestrial, 214, 216
  RCA and, 199
  regulatory policy, 9–11
  ultrahigh frequency, 208
  very high frequency, 208
Telex system, 100, 186
Telpak tariffs, 175
TGV, 98
The Computerization of Society (Minc and Nora), 146–147
The Deal of the Century (Coll), 174
The Economics of Regulation (Kahn), 168
The End of the American Century (Schlosstein), 16
The Information Society (1972 report), 15
The Myth of America's Decline (Nau), 16
The National Information Infrastructures, 22
The Reckoning (Halberstam), 16
"Third Computer Inquiry," 109–110
Thomson (electronics firm), 148–149, 213
Ticker tapes, 223
Time Division Multiplexed (TDM) circuits, 178
Time Incorporated, 14
Time-sharing, computer, 185–191, 233
Time-Werner merger, 14
Titanic sinking, 196

Touch-Tone dialing, 180
Trade deficit, United States, 16, 126, 135
Trading Places (Prestowitz), 16
Traditionalists, 16–18
Train à Grande Vitesse (TGV), 98
Tram roads, 95
Transport Control Protocol (TCP), 76
Transportation Act of 1920, 166
Transportation networks, 58–63
  antitrust legislation and, 5
  automobiles, 92
  bicycles, 92
  communications technology and, xv, 59–63
  horse, 46
  privatization and, 57–58
  telecommunications network and, 59–63
Truman administration, 184
Turing, Alan, 41, 71, 102
Turing model, 108
TYMNET, 75

U.S. *See* United States
UHF, 208
ULSI, 104
Ultra large scale integration (ULSI), 104
Ultrahigh frequency (UHF), 208
Unisys, 103
United Fruit, 196, 198
United States Advisory Council on the National Information Infrastructure, 22
United States Naval Intelligence, 229
United States (U.S.)
  economy, xv–xvi, 3, 58, 116, 135–141, 146
  foreign investors in, 136–137
  Internet in, 145
  manufacturing in, 16–17, 128
  technology policy, 135–141
  trade deficit, 16, 126, 135
Univac, 103, 229

Universal access, 29–30, 254–255
Universal service, 165, 204

Vail, Theodore, 157–161, 174, 248
Van Deerlin, Lionel, 182
Vanderbilt, Cornelius, 91
Vanderbilt family, 52–53
Very high frequency (VHF), 208
Very large scale integration (VLSI), 104
VHF, 208
VLSI, 104
von Neumann, John, 225–226

Wages, American, 17, 126
Walker Report of 1938, 168–169
Watson, Thomas, Jr., 225
Watson, Thomas, Sr., 224
WEAF radio station, 197
Weaver, Warren, 70
West Shore Railroad, 54
Western Electric
  air defense systems and, 70–71, 176
  AT&T and, 158, 177
  Consent Decree of 1956, 71, 173–174, 184–185
  microwaves and, 174
  radio and, 198
  time-sharing and, 186–187
  vacuum tubes and, 231
Western Union
  antitrust legislation and, 91–92, 186–187
  AT&T and, 157–159, 164, 250
  Gould and, 52–54
Western Union and Postal Telegraph, 198
Westinghouse, 196
Wheatstone-Cooke telegraph system, 98
Whirlwind computer, 228, 233
Whirlwind real-time computer research, 104
Wickersham, George W., 161, 163–164

WIDE, 145
Wideband continental microwave network, 227
Widely Distributed Environment (WIDE), 145
Willis-Graham Act of 1921, 166, 204
Wilson administration, 204
Wilson, Woodrow, 161, 195
Wireless technology
  digital revolution and, 241
  Open Communications Infrastructure and, 29
  radio, 60–61
  telegraph systems, 60
  telephone systems, 60–61
  wars, 235–243
Witte Report (1987), 15
WNBC radio station, 197
Works Projects Administration (WPA), 93
World Trade Organization (WTO), 126, 134
WPA, 93
WTO, 126, 134

Zaibatsu tradition, 21
Zworykin, Vladimir, 199